T0259725

The Information Retrieval Series　　Volume 40

More information about this series at http://www.springer.com/series/6128

Tetsuya Sakai
Waseda University
Tokyo, Japan

ISSN 1387-5264
The Information Retrieval Series
ISBN 978-981-13-4581-4   ISBN 978-981-13-1199-4   (eBook)
https://doi.org/10.1007/978-981-13-1199-4

© Springer Nature Singapore Pte Ltd. 2018
Softcover re-print of the Hardcover 1st edition 2018
This work is subject to copyright. All rights are reserved by the Publisher, whether the whole or part of the material is concerned, specifically the rights of translation, reprinting, reuse of illustrations, recitation, broadcasting, reproduction on microfilms or in any other physical way, and transmission or information storage and retrieval, electronic adaptation, computer software, or by similar or dissimilar methodology now known or hereafter developed.
The use of general descriptive names, registered names, trademarks, service marks, etc. in this publication does not imply, even in the absence of a specific statement, that such names are exempt from the relevant protective laws and regulations and therefore free for general use.
The publisher, the authors, and the editors are safe to assume that the advice and information in this book are believed to be true and accurate at the date of publication. Neither the publisher nor the authors or the editors give a warranty, express or implied, with respect to the material contained herein or for any errors or omissions that may have been made. The publisher remains neutral with regard to jurisdictional claims in published maps and institutional affiliations.

This Springer imprint is published by the registered company Springer Nature Singapore Pte Ltd.
The registered company address is: 152 Beach Road, #21-01/04 Gateway East, Singapore 189721, Singapore

Tetsuya Sakai

Laboratory Experiments in Information Retrieval

Sample Sizes, Effect Sizes, and Statistical Power

 Springer

Preface

Classical statistical significance tests are useful to some extent in information retrieval (IR) evaluation, if we can regard the data at hand (e.g., a set of search queries or "topics" from a test collection) as a random sample from some population. In such a situation, we usually want to discuss population parameters (e.g., population means) based on what we have observed (e.g., sample means and variances). However, significance tests are sometimes misinterpreted and/or misused. In the first half of this book (Chaps. 1, 2, 3, 4, and 5), I first review parametric significance tests for comparing system means (i.e., t-tests and ANOVAs) and show how easily they can be conducted using Microsoft Excel or R. I also discuss a few multiple comparison procedures for researchers who are interested in comparing every system pair, including a randomised version of the Tukey HSD test. I then discuss known limitations of classical significance testing and provide practical guidelines for reporting research results regarding comparison of means. These chapters (minus Chap. 1 Sect. 1.3 where noncentral distributions are discussed) should be suitable for undergraduate students in computer science.

Chapters 6 and 7 (plus the aforementioned Chap. 1 Sect. 1.3), in which statistical power is discussed, are probably suitable for graduate students and researchers who regularly conduct laboratory experiments. In Chap. 6, I introduce topic set size design, which leverages the statistical power and confidence interval techniques as described in Nagata (see below), to enable test collection builders to determine an appropriate number of topics to create. My Excel tools for topic set size design based on t-tests, one-way ANOVA and confidence intervals are easy to use. In Chap. 7, I describe power-analysis-based methods for determining an appropriate sample size for a new experiment based on a similar experiment done in the past. My R tools for power analysis with t-tests and ANOVA were adopted from those developed by Toyoda (see below), and rely on R libraries including `pwr`. I provide case studies from IR for both Excel-based topic set size design and R-based power analysis.

My main claims in this book are as follows. Researchers should conduct statistically well-designed experiments so that the right amount of effort is spent for seeking the truth; they should report experimental results appropriately so that their collective efforts will actually add up. I discuss IR as the primary target of research, but the techniques discussed in this book are applicable to related fields such as natural language processing and recommendation, as long as data at hand can be regarded as a random sample from some population.

This book occasionally cites work from IR literature in the 1960s and 1970s. Some of them are available for download from the SIGIR Museum.[1]

My five Excel tools for topic set size design (described in Chap. 6) are based on a Japanese book by Professor Yasushi Nagata of Waseda University. He kindly answered my numerous questions about statistical significance testing and sample size design. My five R scripts for power analysis (described in Chap. 7) are based on the original scripts written by Professor Hideki Toyoda, also of Waseda University. He kindly allowed me to modify his scripts and distribute them for research purposes. I thank these two professors for publishing fantastic Japanese books on these topics, from which I have learnt a lot.

I am also very grateful to Dr. Shunsuke Horii of Waseda University for carefully checking my manuscript and giving me very useful comments, and to my anonymous reviewer who gave me excellent suggestions from the perspective of an IR evaluation expert. I would also like to thank Mio Sugino of Springer for her patience and support and my PhD student Zhaohao Zeng for checking the manuscript, including the code provided in this book.

Finally, I thank my family (Miho, Rio, and Sunny-Uni-Hermione) for understanding the life and responsibilities of an academic and supporting me all the time.

Tokyo, Japan   Tetsuya Sakai
June 2018

Contents

1 Preliminaries 1
  1.1 Principles of Significance Testing 2
     1.1.1 Sample Means and Population Means 2
     1.1.2 Hypotheses, Test Statistics, and P-Values.................... 3
     1.1.3 α, β, and Statistical Power 5
  1.2 Well-Known Probability Distributions 6
     1.2.1 Law of Large Numbers 6
     1.2.2 Normal Distribution and the Central Limit Theorem 8
     1.2.3 χ^2 Distribution.. 11
     1.2.4 t Distribution 13
     1.2.5 F Distribution.. 16
  1.3 Less Well-Known Probability Distributions 17
     1.3.1 Noncentral t Distribution.................................... 17
     1.3.2 Noncentral χ^2 Distribution................................. 22
     1.3.3 Noncentral F Distributions.................................. 22
  References 24

2 t-Tests 27
  2.1 Introduction 27
  2.2 Paired t-Test 29
  2.3 Two-Sample t-Test.. 30
  2.4 Welch's Two-Sample t-Test 31
  2.5 Which Two-Sample t-Test?.. 32
  2.6 Conducting a t-Test with Excel 33
  2.7 Conducting a t-Test with R 36
  2.8 Confidence Intervals for the Difference in Population Means 39
  References 41

3 Analysis of Variance 43
  3.1 One-Way ANOVA 44
     3.1.1 One-Way ANOVA with Equal Group Sizes 44
     3.1.2 One-Way ANOVA with Unequal Group Sizes 47

3.2   Two-Way ANOVA Without Replication 49
3.3   Two-Way ANOVA with Replication 51
3.4   Conducting an ANOVA with Excel 53
3.5   Conducting an ANOVA with R.................................... 54
3.6   Confidence Intervals for System Means 56
References 57

4   **Multiple Comparison Procedures** 59
4.1   Introduction 59
4.2   Familywise Error Rate 60
4.3   Bonferroni Correction 61
    4.3.1   Principles and Limitations of the Bonferroni Correction 61
    4.3.2   Bonferroni Correction with R 62
4.4   Tukey HSD Test... 67
    4.4.1   Tukey HSD with Unequal Group Sizes...................... 68
    4.4.2   Tukey HSD with Equal Group Sizes........................ 69
    4.4.3   Tukey HSD with Paired Observations 70
    4.4.4   Simultaneous Confidence Intervals 71
    4.4.5   Tukey HSD with R... 71
4.5   Randomisation Test and Its Tukey HSD Version 73
    4.5.1   Randomisation Test for Two Systems 74
    4.5.2   Randomised Tukey HSD Test 77
References 80

5   **The Correct Ways to Use Significance Tests** 81
5.1   Limitations of Significance Tests................................ 81
    5.1.1   Criticisms from the Literature........................... 81
    5.1.2   Three Problems, Among Others 83
5.2   Effect Sizes 85
    5.2.1   Effect Sizes for t-Tests 85
    5.2.2   Effect Sizes for Tukey HSD and RTHSD..................... 88
    5.2.3   Effect Sizes for ANOVA 89
5.3   How to Report Your Results 92
    5.3.1   Comparing Two Systems 93
    5.3.2   Comparing More Than Two Systems 95
References 97

6   **Topic Set Size Design Using Excel**................................ 99
6.1   Overview of Topic Set Size Design 99
6.2   Topic Set Size Design with the Paired t-Test 102
    6.2.1   How to Use the Paired t-Test-Based Tool 102
    6.2.2   How the Paired t-Test-Based Topic Set Size
             Design Works 103
6.3   Topic Set Size Design with the Two-Sample t-Test 108
    6.3.1   How to Use the Two-Sample t-Test-Based Tool 108

Contents

      6.3.2   How the Two-Sample t-Test-Based Topic Set Size
              Design Works ....
   6.4 Topic Set Size Design with One-Way ANOVA.............
      6.4.1   How to Use the ANOVA-Based Tool ....
      6.4.2   How the ANOVA-Based Topic Set Size Design Wor
   6.5 Topic Set Size Design with Confidence Intervals for Paired
      6.5.1   How to Use the Paired-Data CI-Based Tool ....
      6.5.2   How the Paired-Data CI-Based Tool Works ....
   6.6 Topic Set Size Design with Confidence Intervals
      for Unpaired Data ....
      6.6.1   How to Use the Two-Sample CI-Based Tool ....
      6.6.2   How the Two-Sample CI-Based Tool Works ....
   6.7 Estimating Population Variances ....
   6.8 Comparing the Different Topic Set Size Design Methods ......
      6.8.1   Paired and Two-Sample t-Tests vs. One-Way ANOVA ..
      6.8.2   CI-Based Topic Set Size Design: Paired
              vs. Unpaired Data..
      6.8.3   One-Way ANOVA vs. Confidence Intervals..............
   References ....

7 Power Analysis Using R ....
   7.1 Introduction ....
   7.2 Overview of the R Scripts for Power Analysis ....
   7.3 Power Analysis with the Paired t-Test ....
   7.4 Power Analysis with the Two-Sample t-Test ....
   7.5 Power Analysis with One-Way ANOVA.............................
   7.6 Power Analysis with Two-Way ANOVA Without Replication ......
   7.7 Power Analysis with Two-Way ANOVA with Replication ....
   7.8 Summary ....
   References ....

8 Conclusions ....
   8.1 A Quick Summary of the Book ....
   8.2 Statistical Reform in IR?...
   References ....

Index... 1

Chapter 1
Preliminaries

Abstract This chapter discusses the basic principles of classical statistical significance testing (Sect. 1.1) and defines some well-known probability distributions that are necessary for discussing parametric significance tests (Sect. 1.2). ("A problem is *parametric* if the form of the underlying distribution is known, and it is *nonparametric* if we have no knowledge concerning the distribution(s) from which the observations are drawn." Good (Permutation, parametric, and bootstrap tests of hypothesis, 3rd edn. Springer, New York, 2005, p. 14). For example, the *paired t-test* is a parametric test for paired data as it relies on the assumption that the observed data independently obey normal distributions (See Chap. 2 Sect. 2.2); the *sign test* is a nonparametric test; they may be applied to the same data if the normality assumption is not valid. This book only discusses parametric tests for comparing means, namely, *t*-tests and ANOVAs. See Chap. 2 for a discussion on the robustness of the *t*-test to the normality assumption violation.) As this book is intended for IR researchers such as myself, not statisticians, well-known theorems are presented without proofs; only brief proofs for corollaries are given. In the next two chapters, we shall use these basic theorems and corollaries as black boxes just as programmers utilise standard libraries when writing their own code. This chapter also defines less well-known distributions called *noncentral* distributions (Sect. 1.3), which we shall need for discussing sample size design and power analysis in Chaps. 6 and 7. Hence Sect. 1.3 may be skipped if the reader only wishes to learn about the principles and limitations of significance testing; however, such readers should read up to Chap. 5 before abandoning this book.

Keywords *p*-values · Significance testing · Statistical power · Statistical significance

© Springer Nature Singapore Pte Ltd. 2018   1
T. Sakai, *Laboratory Experiments in Information Retrieval,*
The Information Retrieval Series 40, https://doi.org/10.1007/978-981-13-1199-4_1

1.1 Principles of Significance Testing

1.1.1 Sample Means and Population Means

In laboratory experiments for evaluating two or more systems, we often compare multiple *sample means* to discuss which systems might be better than others. For example, in experimental IR papers, we see tables containing *mean average precision* (MAP) [1] values, mean *normalised discounted cumulative gain* (nDCG) [9] values, etc. all the time. That is, given n topics (or *search requests*) for evaluating the systems, we obtain n *Search Engine Result Pages* (SERPs) from each system, and an nDCG score can be computed for each SERP; the per-topic nDCG scores are then averaged over the n topics, for each system.[1]

I used the term "sample mean" which implies that I regard the n values as a *random sample* from a *population* of possible nDCG scores for a particular system.[2] A random sample means that I draw an nDCG score at random for n trials, completely independently. For example, suppose we have a query log from a web search engine, which contains an infinite number of queries. If we randomly draw $n = 100$ queries, this can probably be regarded as a random sample; if we randomly draw another 100 queries, that would be another random sample, different from the first one. From different samples, we generally obtain different sample means, and therefore we can regard the sample mean for a system as a *realisation* (i.e., actual value taken) of a continuous *random variable*; in this book, we only consider continuous random variables, so "continuous" shall be omitted hereafter. Henceforth, we primarily consider a set of evaluation measure scores (such as nDCG scores) as a sample.[3]

Figure 1.1 depicts a typical situation where two sample means are compared from the viewpoint of classical statistics. We have a set of (say) nDCG scores for System X, which are realisations[4] of a random variable x. From the n realisations of x, as shown on the right, we compute the sample mean $\bar{x}$, which is also a random variable. Here, we assume that there is an underlying *population distribution* whose *population mean* is μ_X, a constant. The curve (whatever it actually looks like) is

---

[1]Henceforth, for simplicity, I will basically ignore the difference between "topics" (i.e., information need statements) and "queries" (i.e., the sets of keywords input to the search engine) and that between "queries" and the "scores" computed for their search results.

[2]While this book primarily discusses sample means over n topics, some IR researchers have explored the approach of regarding the *document collection* used in an experiment as a sample from a large population of documents [4, 14, 15].

[3]Is a TREC (Text REtrieval Conference) topic set a random sample? Probably not. However, the reader should be aware that IR researchers who rely on significance tests such as t-tests and ANOVA for comparing system means implicitly rely on the basic assumption that a topic set is a random sample. The exact assumptions for t-tests and ANOVA are stated in Chaps. 2 and 3. Also, a computer-based significance test that does not rely on the random sampling and any distributional assumptions will be described in Chap. 4 Sect. 4.5.

[4]In this book, a random variable and its realisations are denoted by the same symbol, e.g. x.

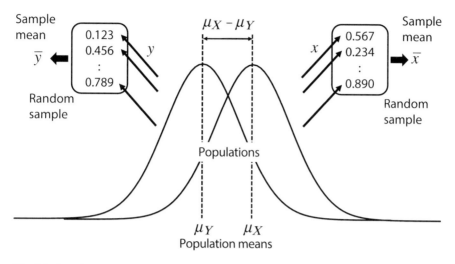

Fig. 1.1 Sample means and population means

an example of a *probability density function* (pdf), which returns a nonnegative value for any input so that the area under the curve adds up to 1. In the figure, the population mean μ_X is the value that is most likely to occur, while values towards the end of the two tails (i.e., those far from μ_X) are much less likely to occur. Similarly, the nDCG scores for System Y are regarded as realisations of a random variable y, from which we obtain the sample mean $\bar{y}$ (another random variable), where the underlying population μ_Y is a constant.

Thus, from data, we obtain sample means as realisations of random variables; they vary, whereas population means are constants.[5] The question we often want to ask is: "What do these sample means tell us about the population means?" For example, what can we say about the "real" difference between Systems X and Y, measured as $\mu_X - \mu_Y$, given the two samples shown in Fig. 1.1? If you are an IR researcher, System X could be your proposed system, and System Y could be a system described in prior art; you probably wish that the population mean μ_X of (say) nDCG scores is actually greater than μ_Y.

1.1.2 Hypotheses, Test Statistics, and P-Values

As was mentioned above, suppose that you are interested in whether the population means μ_X and μ_Y are really different, based on the observation that your sample

---

[5]In contrast to classical significance testing, *Bayesian statistics* [2, 10, 17–19] treat population parameters as random variables. See also Chap. 8.

means $\bar{x}$ and $\bar{y}$ are different. To answer your question, in classical significance testing, we set up a *null hypothesis* H_0, which says that $\mu_X = \mu_Y$. That is, H_0 says that the two population means are actually the same and that the difference you observed in the sample means just reflects the properties of that particular sample. In this section, although we continue with our specific example of comparing two means, note that what is actually presented here is the general principle of classical significance testing.

Along with H_0, we also consider an *alternative hypothesis* H_1: in a *two-sided* test,[6] the alternative to H_0 is $H_1, \mu_X \neq \mu_Y$; in a *one-sided* test, the alternative is $H_1, \mu_X > \mu_Y$; or $H_1, \mu_X < \mu_Y$, depending on your belief or your need. Throughout this book, I shall focus on two-sided t-tests, which are more *conservative* (i.e. reluctant to yield a statistically significant difference) than their one-sided counterparts.[7] IR researchers often hope that H_1 is true that the two systems are *really* different in terms of retrieval effectiveness.

Having set up H_0 and H_1, we then compute a *test statistic* from the observed data: proven theorems from statistics tell us that, under a few basic assumptions (e.g. the samples are drawn independently from normal distributions) *plus the assumption that H_0 is true*, this test statistic obeys a well-understood distribution (e.g. t distribution, F distribution, etc.), which is referred to as the *null distribution*. Figure 1.2 shows an example of a pdf of a null distribution and a test statistic. Recall that the area under the curve adds up to one; the two purple tails of the curve add up to what is known as the *p-value*, defined as the probability of observing the test statistic or something more extreme *given that H_0 is true*. It is worth re-emphasising that we do *not* know whether the test statistic *actually* obeys this distribution; we only know that it should obey this distribution *if H_0 is true*.

Prior to actually examining the data, we decide on a *significance criterion* α, for the purpose of either accepting or rejecting H_0, i.e., to make a dichotomous decision.[8] Let us consider $\alpha = 0.05$ and look at Fig. 1.2 again: the two red tails of the curve each represents a 2.5% of the area under the curve and add up to 5%. The point on the x axis that defines the 2.5% area on the right is called the (two-sided) *critical value* for α. In this particular example, it can be observed that the p-value (sum of the purple tails) is smaller than α (sum of the red tails) or, equivalently, that

---

[6]Not to be confused with *two-sample* tests, which means you have a sample for System X and a different sample for System Y, possibly with different sample sizes (See Chap. 2 Sect. 2.3).

[7]A one-sided test of the form $H_1 : \mu_X > \mu_Y$ would make sense if either $\mu_X < \mu_Y$ is simply impossible, or if you do not need to consider the case where $\mu_X < \mu_Y$ even if this is possible [12]. For example, if you are measuring the effect of introducing an aggressive stemming algorithm into your IR system in terms of recall (not precision) and you *know* that this can never hurt recall, a one-side test may be appropriate. But in practice, when you propose a new IR algorithm and want to compare it with a competitive baseline, it is rarely the case that you know in advance that your proposal is better. Hence I recommend the two-sided test as default. Whichever you choose, hypotheses H_0 and H_1 must be set up *before* actually examining the data.

[8]"Dichotomous thinking," one of the major reasons why classical significance testing has been severely criticised for decades, will be discussed in Chap. 5.

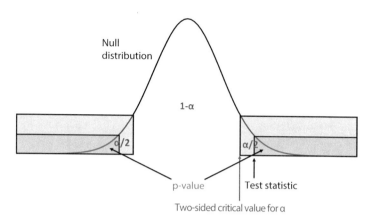

Fig. 1.2 Null distribution, test statistic, p-value, and the significance criterion α

the test statistic is larger than the critical value. Assuming that the test statistic really obeys the null distribution, there is less than 5% chance that we happen to obtain such an extreme value. How come we observed such an unlikely event?

Whenever we observe a test statistic that is less than $100\alpha\%$ likely according to the null distribution, we blame the null hypothesis H_0: we have assumed that H_0 is true and therefore that the test statistic obeys the null distribution, but since it does *not* seem to obey the null distribution (with $100(1 - \alpha)\%$ confidence, since this is not totally impossible), H_0 is probably wrong. Hence we *reject H_0*. For example, if H_0 said that the two population means are equal, we conclude that the difference between the two systems is *statistically significant* at (say) $\alpha = 0.05$.

What if the p-value is larger than α or, equivalently, the test statistic is smaller than the critical value for α in Fig. 1.2? In this case, we *cannot reject H_0*; all we know is that the test statistic lies near the middle of the curve and does not contradict with H_0. In this case, we say that the difference is *not* statistically significant at $\alpha = 0.05$. Note that such a significance test result does *not* mean that the two systems are actually equivalent; it just means "we cannot tell from the data."

1.1.3 α, β, and Statistical Power

Table 1.1 shows the relationship between the truth (i.e. whether H_0 is true or false) and our decisions (i.e. whether to reject H_0 or not). If H_0 is true, then Fig. 1.2 is correct, that is, the test statistic actually obeys the null distribution. From the upper row of Table 1.1, it is clear that when H_0 is true, we manage not to reject it $100(1 - \alpha)\%$ of the time, while rejecting it $100\alpha\%$ of the time. Thus, by construction, we make a *Type I error* with $100\alpha\%$ probability. Alternatively, if H_0 is false, we might make a *Type II error*, which means "failing to reject an incorrect null hypothesis," with some probability β. Ideally, we would like to correctly reject an incorrect H_0;

Table 1.1 α, β and statistical power

	We cannot reject H_0 (not statistically significant)	We reject H_0 (statistically significant)
H_0 is true (e.g. systems are equivalent)	Correct conclusion (probability: $1 - \alpha$)	Type I error (probability: α)
H_0 is false (e.g. systems are not equivalent)	Type II error (probability: β)	Correct conclusion (probability: $1 - \beta$)

that is, we would like to detect a real difference whenever there is one. Clearly, the probability of reaching this type of correct conclusion is $100(1 - \beta)\%$, which is known as the *statistical power* of an experiment [11].

Recall from Sect. 1.1.2 that it is easy to set α: all we have to do is to declare in advance what α is going to be and then compute a p-value from data and reject H_0 if the p-value is smaller than α. In contrast, ensuring that β (or, equivalently, the power) takes a particular value is not as straightforward, because of the relationships among α, β, the sample size, and the *effect size* [3, 5, 8], which will be discussed extensively in Chaps. 5, 6, and 7. At any rate, a well-designed experiment should have a reasonably high statistical power $(1 - \beta)$: what is the point of conducting an experiment if we are highly likely to miss differences that are actually present? In practice, a high statistical power can be ensured by choosing a large enough sample size. The desired Type I and Type II error probabilities α and β should be set according to the actual needs of researchers, but a typical setting is *Cohen's five-eighty convention* [6], which represents $\alpha = 0.05$ and $\beta = 0.20$ (i.e. 80% power). This means that a Type I error is taken to be four times as serious as a Type II error. While this does *not* mean that IR researchers should follow it blindly, henceforth we shall primarily consider this particular setting by default.

1.2 Well-Known Probability Distributions

1.2.1 Law of Large Numbers

Consider a random variable x whose probability density function (pdf) is given by $f(x)$, which tells us how likely a particular realisation of x will occur. Recall that $f(x) \geq 0$ for any x and that $\int_{-\infty}^{\infty} f(x)dx = 1$ (See Sect. 1.1.1). The *expectation* of any given function $g(x)$ is defined as:

$$E(g(x)) = \int_{-\infty}^{\infty} g(x)f(x)dx . \tag{1.1}$$

In particular, $E(x)$, i.e., the expectation of $g(x) = x$, is a constant called the *population mean*. This is the "central" position of x as it is observed an infinite number of times. Moreover, consider $g(x) = (x - E(x))^2$: we call $V(x) = E((x - E(x))^2)$ the *population variance* of x, which is a constant that quantifies how the random variable x varies from the population mean. The *population standard deviation* of x is defined as $\sqrt{V(x)}$.

Before introducing the *Law of Large Numbers*, let us formally define the notion of *independence*. Consider random variables x and y and a *joint probability density function* $f(x, y)$, which tells us how likely a realisation of the combination (x, y) occurs at the same time. As before, $f(x, y)$ should have the following properties:

$$f(x, y) \geq 0, \quad \int_{-\infty}^{\infty} f(x, y)\mathrm{d}x\mathrm{d}y = 1 . \tag{1.2}$$

Also, consider the following *marginal probability density functions*:

$$f_x(x) = \int_{-\infty}^{\infty} f(x, y)\mathrm{d}y , \quad f_y(y) = \int_{-\infty}^{\infty} f(x, y)\mathrm{d}x . \tag{1.3}$$

Thus, $f_x(x)$ has eliminated the random variable y from consideration, and $f_y(y)$ has eliminated x from consideration.

If the following holds for any realisations of (x, y), then x and y are said to be *independent*:

$$f(x, y) = f_x(x) f_y(y) . \tag{1.4}$$

Informally, "x and y do not influence each other." Independence for more than two random variables can be defined in a similar manner.

Theorem 1 (Law of large numbers) *Let $x_1, x_2, \ldots, x_n$ be independent and identically distributed random variables such that $E(x_j) = \mu$ (and $V(x_j) = \sigma^2$) for $j = 1, 2, \ldots, n$. Then, as we increase the sample size n, the sample mean given by*

$$\bar{x} = \frac{1}{n} \sum_{j=1}^{n} x_j \tag{1.5}$$

converges[9] to the population mean μ [12].

Note that Theorem 1 does not say anything about what the common distribution actually is. This theorem says that the population mean can be estimated more and more accurately from the sample mean by just increasing the sample size. Hence,

---

[9]For a discussion on the difference between *convergence in probability* (which is used in the *weak* law of large numbers) and *almost sure convergence* (which is used in the *strong* law of large numbers), see, for example, https://stats.stackexchange.com/questions/2230/convergence-in-probability-vs-almost-sure-convergence.

if a researcher is interested in the population mean of (say) nDCG scores for an IR system, it would be better to estimate it using a sample mean over a large topic set than one over a small topic set.

1.2.2 Normal Distribution and the Central Limit Theorem

The *normal distribution*, one of the most fundamental distributions used in significance testing, is formally defined as follows. For a given pair of parameters μ and $\sigma (> 0)$, the pdf of a normal distribution is given by:

$$f(x; \mu, \sigma) = \frac{1}{\sqrt{2\pi}\sigma} \exp\left\{-\frac{(x-\mu)^2}{2\sigma^2}\right\} \quad (-\infty < x < \infty). \tag{1.6}$$

From the above formula, note that a normal distribution is symmetric across μ; it is known that the population mean and the population variance of the above normal distribution are exactly μ and σ^2, respectively:

$$E(x) = \mu, \tag{1.7}$$

$$V(x) = \sigma^2. \tag{1.8}$$

We denote such a normal distribution by $N(\mu, \sigma^2)$.

The normal distribution has several convenient properties besides the ones mentioned above. Let us discuss a few of them below.

Theorem 2 (Standardisation) *If x obeys $N(\mu, \sigma^2)$, then $u = \frac{x-\mu}{\sigma}$ obeys $N(0, 1^2)$* [13].

Here, $N(0, 1^2)$, that is, a normal distribution with mean 0 and standard deviation 1, is called a *standard normal distribution*.

Let u be a random variable that obeys $N(0, 1^2)$. Then, for any given probability P, the value $z_{inv}(P)$ that satisfies $Pr\{u \geq z_{inv}(P)\} = P$ is called the *upper $100P\%$ z-value*.[10] That is, on the pdf of a standard normal distribution, the area under the curve *to the right of* $z_{inv}(P)$ is exactly P. Owing to the symmetry of the pdf across zero, the following holds:

$$Pr\{u \geq z_{inv}(P)\} = Pr\{u \leq z_{inv}(1 - P)\} = P. \tag{1.9}$$

For example, the upper 5% of the area under the pdf curve ($Pr\{u \geq z_{inv}(0.05)\}$) is equal to the area *to the left of* an upper 95% z-value ($Pr\{u \leq z_{inv}(0.95)\}$).

---

[10]With Microsoft Excel, $z_{inv}(P)$ can be obtained as NORM.S.INV(1 − P); with R, it can be obtained as qnorm(P, lower.tail=FALSE).

Theorem 3 *If x obeys $N(\mu, \sigma^2)$, then for any constants a and b, $ax + b$ obeys $N(a\mu + b, a^2\sigma^2)$* [13].

Theorem 4 (Reproductive property)   *Let $a_1, a_2, \ldots, a_n$ be constants. If $x_1, x_2, \ldots, x_n$ are independent and obey $N(\mu_j, \sigma_j^2)$, respectively ($j = 1, 2, \ldots, n$), then $a_1 x_1 + a_2 x_2 + \cdots + a_n x_n$ obeys $N(a_1\mu_1 + a_2\mu_2 + \cdots + a_n\mu_n, \ a_1^2\sigma_1^2 + a_2^2\sigma_2^2 + \cdots + a_n^2\sigma_n^2)$* [13].

The above theorem means that "a linear combination of normally distributed random variables still obeys a normal distribution."

Corollary 1 *If $x_1, x_2, \ldots, x_n$ are independent and obey an identical normal distribution $N(\mu, \sigma^2)$, then the sample mean $\bar{x} = \frac{1}{n} \sum_{j=1}^{n} x_j$ obeys $N(\mu, \frac{\sigma^2}{n})$.*

Proof This is a corollary of Theorem 4. In Theorem 4, consider the case where $a_j = 1/n, \mu_j = \mu, \sigma_j = \sigma$. Then:

$$a_1\mu_1 + a_2\mu_2 + \cdots + a_n\mu_n = \mu/n + \mu/n + \cdots + \mu/n = \mu \,, \qquad (1.10)$$

$$a_1^2\sigma_1^2 + a_2^2\sigma_2^2 + \cdots + a_n^2\sigma_n^2 = \sigma^2/n^2 + \sigma^2/n^2 + \cdots + \sigma^2/n^2 = \sigma^2/n \,. \quad (1.11)$$

Corollary 2 *If $x_1, x_2, \ldots, x_n$ are independent and obey an identical normal distribution $N(\mu, \sigma^2)$, the following random variable (standardised sample mean) obeys $N(0, 1^2)$:*

$$u = \frac{\bar{x} - \mu}{\sqrt{\sigma^2/n}} \,. \qquad (1.12)$$

Proof This is a corollary from Corollary 1 and Theorem 2.

   Corollary 2 is important, because it implies that *if* (say) nDCG score differences all obey $N(\mu, \sigma^2)$, then the above u, which is a standardised form of the sample mean $\bar{x}$, obeys $N(0, 1^2)$. However, for IR researchers, the problem is that we do not know the population parameters μ and σ^2. Of these two, μ is not really a problem because as soon as we set up a null hypothesis of the form $\mu = 0$, it disappears from Eq. 1.12. Hence the problem is how to estimate σ^2 and what happens to the property of u if we replace it with that estimate. We shall come back to this point after discussing Corollary 6.

Theorem 5 (Central Limit Theorem)   *Let $x_1, x_2, \ldots, x_n$ be independent and identically distributed random variables such that $E(x_j) = \mu$ and $V(x_j) = \sigma^2$ for $j = 1, 2, \ldots, n$. Then, for a sufficiently large n, the sample mean $\bar{x} = \frac{1}{n} \sum_{j=1}^{n} x_j$ approximately obeys $N(\mu, \frac{\sigma^2}{n})$* [12].

Note the difference between Corollary 1 and this theorem; here, there is no requirement that the individual random variables be normally distributed. The central limit theorem is indeed *central* to classical significance testing, for it says that the sample mean with a "sufficiently large" n can be treated as normally distributed *even if the original population distribution is not normal*. Consider nDCG or any other measure that has the [0, 1] range: such scores are clearly *not* normally distributed since the range of any normal distribution is $(-\infty, \infty)$. However, the theorem says that, given a "sufficiently large" topic set, the *mean* of the scores can be regarded as a random variable that obeys a normal distribution.

How large is "sufficiently large" then? Of course, the answer depends on what the original distribution looks like, but even a sample size as small as $n = 20$ may suffice for many practical applications. For example, if we draw many samples of size $n = 5$ from a *uniform* distribution, the distribution of $\bar{x}$ already looks like a normal distribution; even for a *binomial distribution* which only gives us either 0 or 1 with probability $P = 0.5$, if we draw many samples of size $n = 15$ from it, the distribution of $\bar{x}$ already looks quite like a normal distribution. However, if the probability for the binomial distribution is (say) $P = 0.2$, the distribution of the samples of size $n = 15$ still looks skewed [12].[11]

Corollary 3 (An Alternative Form of the Central Limit Theorem.) *Let* $x_1, x_2, \ldots, x_n$ *be independent and identically distributed random variables such that* $E(x_j) = \mu$ *and* $V(x_j) = \sigma^2$ *for* $j = 1, 2, \ldots, n$. *Moreover, let* $\bar{x} = \frac{1}{n} \sum_{j=1}^{n} x_j$. *Then, for a sufficiently large n,* $u = \frac{\bar{x} - \mu}{\sigma}$ *approximately obeys* $N(0, 1^2)$.

Proof This is a collorary from Theorems 2 and 5.

If $\bar{x}$ obeys $N(\mu, \frac{\sigma^2}{n})$, then, from Eq. 1.7, $E(\bar{x}) = \mu$ holds: we say that the sample mean $\bar{x}$ is an *unbiased estimator* of the population mean μ. Below, we shall discuss an unbiased estimator of the population *variance*. Unbiased estimators are useful, because we never actually know the population means and variances, and therefore we often need to estimate them as accurately as possible from the samples that we have.

Recall that for a random variable x, the population variance is given by $V(x) = E((x - E(x))^2)$. To compute something similar from a sample, let us consider a *sum of squares*:

$$S = \sum_{j=1}^{n} (x_j - \bar{x})^2 = \sum_{j=1}^{n} x_j^2 - \frac{(\sum_{j=1}^{n} x_j)^2}{n} . \qquad (1.13)$$

Furthermore, let the *sample variance* be:

---

[11] It is often recommended that a binomial distribution be approximated with a normal distribution provided $nP \geq 5$ *and* $n(1 - P) \geq 5$ hold [12]. Note that when $n = 15$ *and* $P = 0.5$, we have $nP = n(1 - P) = 7.5 > 5$, whereas, when $n = 15$ *and* $P = 0.2$, we have $nP = 3 < 5$. In the latter situation, the above recommendation suggests that we should increase the sample size to $n = 25$: however, note that even this is not a large number.

$$V = \frac{S}{n-1} .$$
(1.14)

Note[12] that the denominator is $n-1$ rather than n. We have the following theorem for V.

Theorem 6 *If* $x_1, x_2, \ldots, x_n$ *are independent and all obey* $N(\mu, \sigma^2)$, *then* $E(V) = \sigma^2$ *holds* [12].

In other words, the sample variance V is an unbiased estimator of the population variance σ^2. This suggests a solution to the aforementioned problem regarding the lack of knowledge about σ^2 in Eq. 1.12: if we replace σ^2 (a constant) with V (a random variable) in Eq. 1.12, u may no longer strictly obey $N(0,1)$ but obey a distribution reasonably similar to it. This is exactly what a t-distribution is, which we rely on when we conduct t-tests.

While the sample variance may also be defined by using n as the denominator rather than $n-1$, that version is *not* an unbiased estimator of σ^2 and is not very useful in the context of classical significance testing.[13]

1.2.3 χ^2 *Distribution*

Here we define the χ^2 *distribution*. In this book, we shall use it to define the t distribution (which we need to conduct t-tests) and the F distribution (which we need to conduct ANOVAs).

Let $u_1, u_2, \ldots, u_k$ be independent random variables that each obey $N(0, 1^2)$. The probability distribution of the following random variable χ^2 is called a χ^2 distribution with $\phi = k$ *degrees of freedom*:

$$\chi^2 = u_1^2 + u_2^2 + \cdots + u_k^2 = \sum_{j=1}^{k} u_j^2 ,$$
(1.15)

---

[12]The sample mean uses n as the denominator because it is based on n independent pieces of information, namely, x_i. In contrast, while Eq. 1.13 shows that S is based on $(x_i - \bar{x})$'s, these are not actually n independent pieces of information, since $\sum_{i=1}^{n}(x_i - \bar{x}) = \sum_i^n x_i - n\bar{x} = 0$ holds. There are only $(n-1)$ independent pieces of information in the sense that, once we have decided on $(n-1)$ values out of n, the last one is automatically determined due to the above constraint. For this reason, dividing S by $n-1$ makes sense [12]. See also Sect. 1.2.4 where the *degrees of freedom* of S is discussed.

[13]To be more specific, $E(S/n) = \frac{n-1}{n}\sigma^2 < \sigma^2$ and hence S/n *underestimates* σ^2 [12]. Also of note is that the *sample standard deviation* $\sqrt{V}$ is *not* an unbiased estimator of the population standard deviation σ despite the fact that $E(V) = \sigma^2$.

and is denoted by $\chi^2(\phi)$. That is, if you construct a sum of squares from independent variables that each obey a standard normal distribution, the distribution of that sum of squares is called a χ^2 distribution. The degrees of freedom ϕ can be interpreted as a measure of how "accurate" the sum of squares is: we shall come back to this point in Sect. 1.2.4.

For any $x > 0$, the pdf of $\chi^2(\phi)$ is given by [13][14]:

$$f(x; \phi) = \frac{1}{\Gamma(\phi/2)2^{\phi/2}} x^{(\phi/2)-1} e^{-x/2} , \tag{1.16}$$

where $\Gamma(a)$ is the gamma function given by:

$$\Gamma(a) = \int_0^\infty x^{a-1} e^{-x} dx \qquad (a > 0) . \tag{1.17}$$

Let χ^2 be a random variable that obeys $\chi^2(\phi)$. Then, for any given probability P, the value $\chi^2_{inv}(\phi; P)$ that satisfies $Pr\{\chi^2 \geq \chi^2_{inv}(\phi; P)\} = P$ is called the *upper* $100P\%$ χ^2-*value*.[15] That is, on the pdf of a χ^2 distribution, the area under the curve *to the right of* $\chi^2_{inv}(\phi; P)$ *is exactly* P.

Corollary 4 *If $x_1, x_2, \ldots, x_n$ are independent and each obey $N(\mu, \sigma^2)$, then the following random variable obeys $\chi^2(n)$:*

$$\frac{\sum_{j=1}^n (x_j - \mu)^2}{\sigma^2} . \tag{1.18}$$

Proof From Theorem 2, $u_j = \frac{x_j - \mu}{\sigma}$ for $j = 1, 2, \ldots, n$ are independent, and each obey $N(0, 1^2)$. Hence, from the definition of χ^2 distribution, the above Corollary holds.

The following theorem is used to derive a property of the t-distribution concerning the two-sample t-test (Sect. 1.2.4) and to discuss how ANOVA works (Chap. 3 Sect. 3.1).

Theorem 7 (Reproductive property) *Consider a random variable χ_1^2 that obeys $\chi^2(\phi_1)$ and another χ_2^2 that obeys $\chi^2(\phi_2)$. If the two are independent, then $\chi_1^2 + \chi_2^2$ obeys $\chi^2(\phi_1 + \phi_2)$* [13].

Again, let $x_1, x_2, \ldots, x_n$ be independent and each obey $N(\mu, \sigma^2)$, and recall the sample mean $\bar{x}$ and the sum of squares S:

---

[14]It is known that the population mean and the population variance of χ^2 are given by $E(\chi^2) = \phi$ and $V(\chi^2) = 2\phi$, respectively.

[15]With Microsoft Excel, $\chi^2_{inv}(\phi; P)$ can be obtained as `CHISQ.INV.RT(P, φ)`; with R, it can be obtained as `qchisq(P, φ, lower.tail=FALSE)`.

$$\bar{x} = \frac{1}{n}\sum_{j=1}^{n} x_j , \quad S = \sum_{j=1}^{n}(x_j - \bar{x})^2 . \tag{1.19}$$

We will also need the following two theorems concerning S to discuss an important property of the t distribution in Sect. 1.2.4.

Theorem 8 $\bar{x}$ *and S are independent* [13].

Theorem 9 S/σ^2 *obeys* $\chi^2(n-1)$ [13].

Corollary 4 is used in the proof of Theorem 9. Note that while the random variable discussed in Corollary 4 involves the population mean μ, that discussed in Theorem 9 involves the sample mean $\bar{x}$, and that the degrees of freedom for the latter is $n-1$ rather than n.

1.2.4   t Distribution

Having gone through normal and χ^2 distributions, we can now define the t *distribution*. Let u be a random variable that obeys $N(0, 1^2)$ and χ^2 be a random variable that obeys $\chi^2(\phi)$. Furthermore, let u and χ^2 be independent. Then, the probability distribution of the following random variable t is called a t distribution with ϕ degrees of freedom:

$$t = \frac{u}{\sqrt{\chi^2/\phi}} . \tag{1.20}$$

The aforementioned distribution is denoted by $t(\phi)$.

For any x, the pdf of $t(\phi)$ is given by $[13]^{16}$:

$$f(x; \phi) = \frac{1}{\sqrt{\phi}B(1/2, \phi/2)} \left(1 + \frac{x^2}{\phi}\right)^{-(\phi+1)/2} , \tag{1.21}$$

where $B(a, b)$ is the beta function given by:

$$B(a, b) = \int_0^1 x^{a-1}(1-x)^{b-1}dx \quad (a > 0, \ b > 0) . \tag{1.22}$$

Let t be a random variable that obeys $t(\phi)$. Then, for any given probability P, the value $t_{inv}(\phi; P)$ that satisfies $Pr\{|t| \geq t_{inv}(\phi; P)\} = P$ is called the *two-*

---

[16]It is known that the population mean and the population variance of t are given by $E(t) = 0$ (for $\phi \geq 2$) and $V(t) = \frac{\phi}{\phi-2}$ (for $\phi \geq 3$), respectively.

sided $100P\%$ *t-value.*[17] Note that, while $z_{inv}(P)$ was defined as the *upper* $100P\%$ value, $t_{inv}(\phi; P)$ is defined as a *two-sided* $100P\%$ value, following Nagata [13]: As Fig. 1.2 (Sect. 1.1.2 of this chapter) suggests, a two-sided (say) 5% t-value secures 2.5% on either side of the pdf curve. Owing to the symmetry of the pdf across zero, the following holds:

$$Pr\{t \geq t_{inv}(\phi; P)\} = Pr\{t \leq -t_{inv}(\phi; P)\} = \frac{P}{2} . \tag{1.23}$$

The following properties of the t distribution enable us to conduct t-tests. To be more specific, Corollary 5 forms the basis of the paired t-test (Chap. 2 Sect. 2.2), while Corollary 6 forms the basis of the two-sample t-test (Chap. 2 Sect. 2.3). The paired t-test can be used, for example, when we want to compare two systems with the same topic set; the two-sample t-test can be used, for example, when we want to compare two systems based on a user study where User Group A was assigned to one system and User Group B was assigned to the other.

Corollary 5 *If $x_1, x_2, \ldots, x_n$ are independent and each obey $N(\mu, \sigma^2)$, then the following random variable obeys $t(n-1)$:*

$$t = \frac{\bar{x} - \mu}{\sqrt{V/n}} , \tag{1.24}$$

where $\bar{x}$ is the sample mean and $V = S/(n-1)$ is the sample variance (See Eq. 1.19).

Proof First, from Corollary 2, $u = (\bar{x} - \mu)/\sqrt{\sigma^2/n}$ obeys $N(0, 1^2)$. Second, from Theorem 9, $\chi^2 = S/\sigma^2 = (n-1)V/\sigma^2$ obeys $\chi^2(n-1)$. Therefore,

$$t = \frac{\bar{x} - \mu}{\sqrt{V/n}} = \frac{u\sqrt{\sigma^2/n}}{\sqrt{\chi^2\sigma^2/(n-1)n}} = \frac{u}{\sqrt{\chi^2/(n-1)}} . \tag{1.25}$$

Moreover, from Theorem 8, $\bar{x}$ and S are independent, and therefore u and χ^2 are also independent. Hence, from the definition of t distribution, the above corollary holds.

In the above discussion, the degrees of freedom for the t distribution was $\phi = n-1$. Note that:

$$V = \frac{S}{\phi} . \tag{1.26}$$

---

[17]With Microsoft Excel, $t_{inv}(\phi; P)$ can be obtained as T.INV.2T(P, ϕ); with R, it can be obtained as qt$(P/2, \phi,$ lower.tail=FALSE$)$.

We have already seen when V was first introduced in Eq. 1.14 that V relies on $\phi = n - 1$ independent pieces of information: ϕ represents the degrees of freedom associated with the sum of squares S. Now, compare u given in the above proof, which obeys $N(0, 1^2)$, and t as defined in Eq. 1.24. It can be observed that t is obtained by replacing the population variance σ^2 in the definition of u with a sample variance V; as a result t obeys a distribution very similar to $N(0, 1^2)$. How similar depends on the accuracy of V as an estimator of σ^2, which is exactly what the degrees of freedom ϕ represents. The accuracy can be improved arbitrarily by increasing the sample size n; when $\phi = \infty$, the t distribution is exactly $N(0, 1^2)$.

Corollary 6 *If* $x_{11}, x_{12}, \ldots, x_{1n_1}$ *each obey* $N(\mu_1, \sigma^2)$, $x_{21}, x_{22}, \ldots, x_{2n_2}$ *each obey* $N(\mu_2, \sigma^2)$, *and they are all independent, then the following random variable obeys* $t(n_1 + n_2 - 2)$:

$$t = \frac{\bar{x}_{1\bullet} - \bar{x}_{2\bullet} - (\mu_1 - \mu_2)}{\sqrt{V_p(1/n_1 + 1/n_2)}} , \tag{1.27}$$

where

$$\bar{x}_{1\bullet} = \frac{\sum_{j=1}^{n_1} x_{1j}}{n_1} , \qquad \bar{x}_{2\bullet} = \frac{\sum_{j=1}^{n_2} x_{2j}}{n_2} , \tag{1.28}$$

and V_p *is a* pooled variance *given by* [12]:

$$S_1 = \sum_{j=1}^{n_1} (x_{1j} - \bar{x}_{1\bullet})^2 , \qquad S_2 = \sum_{j=1}^{n_2} (x_{2j} - \bar{x}_{2\bullet})^2 , \qquad V_p = \frac{S_1 + S_2}{n_1 + n_2 - 2} . \tag{1.29}$$

Proof From Corollary 1, $\bar{x}_{1\bullet}$ obeys $N(\mu_1, \sigma^2/n_1)$, while $\bar{x}_{2\bullet}$ obeys $N(\mu_2, \sigma^2/n_2)$. Since $\bar{x}_{1\bullet}$ and $\bar{x}_{1\bullet}$ must also be independent, from Theorem 4, $\bar{x}_{1\bullet} - \bar{x}_{2\bullet}$ obeys $N(\mu_1 - \mu_2, \sigma^2(1/n_1 + 1/n_2))$. Hence, from Theorem 2,

$$u = \frac{\bar{x}_{1\bullet} - \bar{x}_{2\bullet} - (\mu_1 - \mu_2)}{\sqrt{\sigma^2(1/n_1 + 1/n_2)}} \tag{1.30}$$

obeys $N(0, 1^2)$. On the other hand, from Theorem 9, S_1/σ^2 obeys $\chi^2(n_1 - 1)$, while S_2/σ^2 obeys $\chi^2(n_2 - 1)$. Since these must also be independent, from Theorem 7, $\chi^2 = S_1/\sigma^2 + S_2/\sigma^2$ obeys $\chi^2(n_1 + n_2 - 2)$, which gives us $S_1 + S_2 = \chi^2\sigma^2$. Therefore, Eq. 1.27 can be rewritten as:

$$t = \frac{u\sqrt{\sigma^2(1/n_1 + 1/n_2)}}{\sqrt{\frac{\chi^2\sigma^2}{n_1+n_2-2}(1/n_1 + 1/n_2)}} = \frac{u}{\sqrt{\chi^2/(n_1 + n_2 - 2)}} . \tag{1.31}$$

Moreover, from Theorem 8, $\bar{x}_{1\bullet}$ and S_1 are independent, $\bar{x}_{2\bullet}$ and S_2 are independent, and therefore u (Eq. 1.30) and $\chi^2(= (S_1 + S_2)/\sigma^2)$ are independent. Hence, from the definition of t distribution, the above corollary holds.

1.2.5   F Distribution

Let us now define the *F distribution*. Consider a random variable χ_1^2 that obeys $\chi^2(\phi_1)$, and another χ_2^2 that obeys $\chi^2(\phi_2)$. Furthermore, let χ_1^2 and χ_2^2 be independent. Then the probability distribution of the following random variable F is called an F distribution with (ϕ_1, ϕ_2) degrees of freedom:

$$F = \frac{\chi_1^2/\phi_1}{\chi_2^2/\phi_2} \, . \tag{1.32}$$

The aforementioned distribution is denoted by $F(\phi_1, \phi_2)$.

For $x > 0$, the pdf of $F(\phi_1, \phi_2)$ is given by [13][18]:

$$f(x; \phi_1, \phi_2) = \frac{1}{B(\phi_1/2, \phi_2/2)} \left(\frac{\phi_1}{\phi_2}\right)^{\phi_1/2} x^{\phi_1/2-1} \left(1 + \frac{\phi_1 x}{\phi_2}\right)^{-(\phi_1+\phi_2)/2},$$
$$\tag{1.33}$$

where $B(a, b)$ is as defined in Eq. 1.22.

Let F be a random variable that obeys $F(\phi_1, \phi_2)$. Then, for any given probability P, the value $F_{inv}(\phi_1, \phi_2; P)$ that satisfies $Pr\{F \geq F_{inv}(\phi_1, \phi_2; P)\} = P$ is called the *upper* $100P\%$ *F-value*.[19] That is, on the pdf of an F distribution, the area under the curve *to the right of* $F_{inv}(\phi_1, \phi_2; P)$ is exactly P.

The following relationship between the t distribution and the F distribution will be used to show that one-way ANOVA with two groups is strictly equivalent to two-sample t-test in Chap. 6 Sect. 6.8.1.

Corollary 7 *Let t be a random variable that obeys $t(\phi)$. Then, t^2 obeys $F(1, \phi)$.*

Proof From Eq. 1.20, t^2 can be expressed as:

$$t^2 = \frac{u^2}{\chi^2/\phi} = \frac{u^2/1}{\chi^2/\phi} \, , \tag{1.34}$$

---

[18] It is known that the population mean and the population variance of F are given by $E(F) = \frac{\phi_2}{\phi_2-2}$ (for $\phi_2 \geq 3$) and $V(F) = \frac{2\phi_2^2(\phi_1+\phi_2-2)}{\phi_1(\phi_2-2)^2(\phi_2-4)}$ (for $\phi_2 \geq 5$), respectively.

[19] With Microsoft Excel, $F_{inv}(\phi_1, \phi_2; P)$ can be obtained as F.INV.RT(P, ϕ_1, ϕ_2); with R, it can be obtained as qf(P, ϕ_1, ϕ_2, lower.tail=FALSE).

where u obeys $N(0, 1^2)$ and χ^2 obeys $\chi^2(\phi)$. Now, from the definition of the χ^2 distribution (Eq. 1.15), clearly u^2 obeys $\chi^2(1)$. Hence, from the definition of the F distribution (Eq. 1.32), the above corollary holds.

The following property of the F distribution is what enables us to conduct ANOVA, as we shall discuss in Chap. 3.

Theorem 10 *Let* $x_{11}, x_{12}, \ldots, x_{1n_1}$ *be random variables that each obey* $N(\mu_1, \sigma_1^2)$. *Let* $x_{21}, x_{22}, \ldots, x_{2n_2}$ *be random variables that each obey* $N(\mu_2, \sigma_2^2)$. *Moreover, let all of the variables be independent. Furthermore, let:*

$$V_1 = \frac{\sum_{j=1}^{n_1}(x_{1j} - \bar{x}_{1\bullet})^2}{n_1 - 1}, \qquad V_2 = \frac{\sum_{j=1}^{n_2}(x_{2j} - \bar{x}_{2\bullet})^2}{n_2 - 1}, \tag{1.35}$$

where $\bar{x}_{1\bullet}$ *and* $\bar{x}_{2\bullet}$ *are as defined in Eq. 1.28. Then, the following random variable obeys* $F(n_1 - 1, n_2 - 1)$ [13]:

$$\frac{V_1/\sigma_1^2}{V_2/\sigma_2^2}. \tag{1.36}$$

1.3 Less Well-Known Probability Distributions

The noncentral distributions discussed below are required for understanding what goes on behind topic set size design (Chap. 6) and power analysis (Chap. 7). While the *central t* and *F* distributions described in Sect. 1.2 are what the test statistics obey when the null hypothesis H_0 is true, *noncentral t* and *F* distributions described below are the ones they obey when H_0 is false. As was mentioned earlier, this section may be skipped if the reader only wants to learn how to conduct t-tests and ANOVAs and to report the results appropriately.

1.3.1 Noncentral t Distribution

Let y be a random variable that obeys $N(\lambda, 1^2)$ for some constant λ; let χ^2 be a random variable that obeys $\chi^2(\phi)$. If y and χ^2 are independent, the probability distribution of the following random variable t' is called the *noncentral t distribution* with ϕ degrees of freedom and a *noncentrality parameter* λ:

$$t' = \frac{y}{\sqrt{\chi^2/\phi}}, \tag{1.37}$$

and is denoted by $t'(\phi, \lambda)$. Compare Eq. 1.37 with Eq. 1.20: it is clear that when $\lambda = 0$, then $t'(\phi, \lambda)$ reduces to $t(\phi)$. That is, the noncentral t distribution generalises the (central) t distribution.

For the sake of completeness, the pdf of $t'(\phi, \lambda)$ is provided below [13][20]:

$$f(x; \phi, \lambda) = \frac{\exp(-\lambda^2/2)}{\sqrt{\phi \pi} \Gamma(\phi/2)} \sum_{j=0}^{\infty} \Gamma\left(\frac{\phi + j + 1}{2}\right) \frac{(\lambda x)^j}{j!} \left(\frac{2}{\phi}\right)^{j/2} \left(1 + \frac{x^2}{\phi}\right)^{-(\phi+j+1)/2}.$$

$$(1.38)$$

Corollary 8 *If* $x_1, x_2, \ldots, x_n$ *are independent and each obey* $N(\mu, \sigma^2)$, *then the following random variable obeys* $t'(n - 1, \lambda)$:

$$t_0 = \frac{\bar{x} - \mu_0}{\sqrt{V/n}}, \qquad (1.39)$$

where

$$\lambda = \frac{\sqrt{n}(\mu - \mu_0)}{\sigma}. \qquad (1.40)$$

Proof The proof is similar to that of Corollary 5: from that proof, we have $\bar{x} - \mu = u\sigma/\sqrt{n}$ (where u obeys $N(0, 1^2)$) and $V = \chi^2 \sigma^2/(n - 1)$ (where χ^2 obeys $\chi^2(n - 1)$). Also, if we define λ as in Eq. 1.40, then $\mu - \mu_0 = \lambda\sigma/\sqrt{n}$. Therefore,

$$t_0 = \frac{(\bar{x} - \mu) + (\mu - \mu_0)}{\sqrt{V/n}} = \frac{u\sigma/\sqrt{n} + \lambda\sigma/\sqrt{n}}{\sigma\sqrt{\chi^2/(n - 1)n}} = \frac{u + \lambda}{\sqrt{\chi^2/(n - 1)}}. \qquad (1.41)$$

Now, since u obeys $N(0, 1^2)$, $y = u + \lambda$ obeys $N(\lambda, 1^2)$ (Theorem 3). Hence, from the definition of noncentral t distribution, the above corollary holds.

Corollary 9 *If* $x_{11}, x_{12}, \ldots, x_{1n_1}$ *each obey* $N(\mu_1, \sigma^2)$, $x_{21}, x_{22}, \ldots, x_{2n_2}$ *each obey* $N(\mu_2, \sigma^2)$, *and they are all independent, then the following random variable obeys* $t'(n_1 + n_2 - 2, \lambda)$:

$$t_0 = \frac{\bar{x}_{1\bullet} - \bar{x}_{2\bullet}}{\sqrt{V_p(1/n_1 + 1/n_2)}} \qquad (1.42)$$

where

$$\lambda = \frac{\sqrt{n_1 n_2/(n_1 + n_2)}(\mu_1 - \mu_2)}{\sigma}. \qquad (1.43)$$

---

[20]It is known that the population mean and the population variance of t' are given by $E(t') = \frac{\lambda\sqrt{\phi/2}\Gamma((\phi-1)/2)}{\Gamma(\phi/2)}$ (for $\phi \geq 2$) and $V(t') = \frac{\phi(1+\lambda^2)}{\phi-2} - \{E(t')\}^2$ (for $\phi \geq 3$), respectively.

Proof The proof is similar to that of Corollary 6: from that proof, we have

$$\bar{x}_{1\bullet} - \bar{x}_{2\bullet} = u\sigma\sqrt{\frac{n_1 + n_2}{n_1 n_2}} + \mu_1 - \mu_2 \tag{1.44}$$

(where u obeys $N(0, 1^2)$) and $V_p = \chi^2\sigma^2/(n_1 + n_2 - 2)$ (where χ^2 obeys $\chi^2(n_1 + n_2 - 2)$)). Also, if we define λ as in Eq. 1.43, then

$$\mu_1 - \mu_2 = \lambda\sigma\sqrt{\frac{n_1 + n_2}{n_1 n_2}} . \tag{1.45}$$

Therefore,

$$t_0 = \frac{u\sigma\sqrt{\frac{n_1+n_2}{n_1 n_2}} + \lambda\sigma\sqrt{\frac{n_1+n_2}{n_1 n_2}}}{\sqrt{\frac{\chi^2\sigma^2}{n_1+n_2-2}\left(\frac{n_1+n_2}{n_1 n_2}\right)}} = \frac{u + \lambda}{\sqrt{\chi^2/(n_1 + n_2 - 2)}} . \tag{1.46}$$

Now, since u obeys $N(0, 1^2)$, $y = u + \lambda$ obeys $N(\lambda, 1^2)$ (Theorem 3). Hence, from the definition of noncentral t distribution, the above corollary holds.

Theorem 11 (Approximating a noncentral t distribution with a normal distribution) *For random variables t' and u that obey $t'(\phi, \lambda)$ and $N(0, 1^2)$, respectively, the following approximation holds* [13]:

$$Pr\{t' \le w\} \approx Pr\left\{u \le \frac{wc^* - \lambda}{\sqrt{1 + w^2(1 - c^{*2})}}\right\} \tag{1.47}$$

where

$$c^* = E\left(\sqrt{\frac{\chi^2}{\phi}}\right) = \frac{\sqrt{2}\Gamma\left(\frac{\phi+1}{2}\right)}{\sqrt{\phi}\Gamma\left(\frac{\phi}{2}\right)} \tag{1.48}$$

and χ^2 is a random variable that obeys $\chi^2(\phi)$.

A gist of the proof is given in Appendix 1 of Sakai [16].

Corollary 10 *For random variables t' and u that obey $t'(\phi, \lambda)$ and $N(0, 1^2)$, respectively, the following approximation holds* [13]:

$$Pr\{t' \le w\} \approx Pr\left\{u \le \frac{w(1 - 1/4\phi) - \lambda}{\sqrt{1 + w^2/2\phi}}\right\} . \tag{1.49}$$

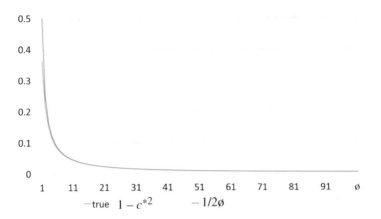

Fig. 1.3 Approximating $1 - c^{*2}$ with $1/2\phi$

Proof First, we approximate Eq. 1.48 by letting $c^* \approx 1 - 1/4\phi$. Then $1 - c^{*2} = (1 + c^*)(1 - c^*) \approx (2 - 1/4\phi)(1/4\phi) \approx 1/2\phi$. Substitute these into Theorem 11, and we have the above corollary.

Figure 1.3 compares the approximation $1 - c^{*2} \approx 1/2\phi$ (orange curve) with its true value computed from Eq. 1.48 (blue curve) for different values of ϕ: it is clear that the approximation is highly accurate.

Corollary 11 *For random variables t' and u that obey $t'(\phi, \lambda)$ and $N(0, 1^2)$, respectively, the following approximation holds:*

$$Pr\{t' \leq w\} \approx Pr\left\{u \leq \frac{w - \lambda}{\sqrt{1 + w^2/2\phi}}\right\}. \tag{1.50}$$

Proof In Corollary 10, further let $1 - 1/4\phi \approx 1$.

Corollary 12 (Approximating a z-value with a t-value) *The following approximation holds:*

$$z_{inv}(P/2) \approx \frac{t_{inv}(\phi; P)}{\sqrt{1 + \{t_{inv}(\phi; P)\}^2/2\phi}}. \tag{1.51}$$

Proof In Corollary 11, let $\lambda = 0$ so that $t = t'$ obeys $t(\phi)$, i.e., a *central t* distribution. By letting $w = t_{inv}(\phi; P)$, we have:

$$Pr\{t \leq t_{inv}(\phi; P)\} \approx Pr\left\{u \leq \frac{t_{inv}(\phi; P)}{\sqrt{1 + t_{inv}(\phi; P)^2/2\phi}}\right\}. \tag{1.52}$$

Hence:

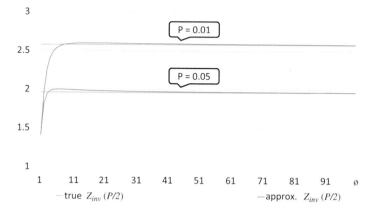

Fig. 1.4 An empirical verification of Corollary 12

$$z_{inv}(P/2) \approx \frac{t_{inv}(\phi; P)}{\sqrt{1 + \{t_{inv}(\phi; P)\}^2/2\phi}} \, . \tag{1.53}$$

Recall that while $z_{inv}(P)$ denotes the *upper* $100P\%$ point on a standard normal distribution (Theorem 2), $t_{inv}(\phi; P)$ denotes the *two-sided* $100P\%$ point on a central t distribution (Sect. 1.1.2). Hence the $P/2$ on the left side.

Figure 1.4 compares the approximations using $t_{inv}(\phi; P)$ (orange curves) with their true values $z_{inv}(P/2)$ (blue curves) for $P = 0.01, 0.05$: it is clear that the approximations are highly accurate except when ϕ is extremely small.

Corollary 13 (Approximating a t-value with a z-value) *The following approximation holds:*

$$t_{inv}(\phi; P) \approx \frac{z_{inv}(P/2)}{\sqrt{1 - \{z_{inv}(P/2)\}^2/2\phi}} \, . \tag{1.54}$$

Proof This can be derived from Corollary 12.

Corollary 14 *The following approximating holds:*

$$1 + t_{inv}(\phi; P)^2/2\phi \approx \frac{1}{1 - z_{inv}(P/2)^2/2\phi} \, . \tag{1.55}$$

Proof From Corollaries 12 and 13:

$$1 + t_{inv}(\phi; P)^2/2\phi \approx \frac{t_{inv}(\phi; P)^2}{z_{inv}(P/2)^2} \approx \frac{\frac{z_{inv}(P/2)^2}{1 - z_{inv}(P/2)^2/2\phi}}{z_{inv}(P/2)^2} = \frac{1}{1 - z_{inv}(P/2)^2/2\phi} \, . \tag{1.56}$$

1.3.2 Noncentral χ^2 Distribution

Let x_i be independent random variables that each obey $N(\mu_i, 1^2)$ ($i = 1, 2, \ldots, n$). The probability distribution of the following random variable χ'^2 is called the *noncentral χ^2 distribution* with $\phi = k$ degrees of freedom and a noncentrality parameter λ:

$$\chi'^2 = x_1^2 + x_2^2 + \cdots + x_k^2 = \sum_{i=1}^{k} x_i^2 , \tag{1.57}$$

and is denoted by $\chi'^2(\phi, \lambda)$. Here, the noncentrality parameter is given by:

$$\lambda = \sum_{i=1}^{k} \mu_i^2 . \tag{1.58}$$

Hence, it is clear that when $\lambda = 0$, x_i each obey $N(0, 1^2)$ and therefore that $\chi'^2(\phi, \lambda)$ reduces to $\chi^2(\phi)$. That is, the noncentral χ^2 distribution generalises the (central) χ^2 distribution.

For the sake of completeness, the pdf of $\chi'^2(\phi, \lambda)$ is provided below [13][21]:

$$f(x; \phi, \lambda) = \sum_{j=0}^{\infty} p(j; \lambda/2)\, g(x; \phi + 2j) , \tag{1.59}$$

where

$$p(j; \lambda/2) = \frac{(\lambda/2)^j}{j!} e^{-\frac{\lambda}{2}} , \tag{1.60}$$

$$g(x; \phi + 2j) = \frac{1}{\Gamma((\phi + 2j)/2)\, 2^{(\phi+2j)/2}}\, x^{(\phi+2j)/2-1} e^{-\frac{x}{2}} . \tag{1.61}$$

1.3.3 Noncentral F Distributions

Let $\chi_1'^2$ be a random variable that obeys $\chi'^2(\phi_1, \lambda)$ (a noncentral χ^2 distribution), and let χ_2^2 be a random variable that obeys $\chi^2(\phi_2)$ (a central χ^2 distribution).

---

[21]It is known that the population mean and the population variance of χ'^2 are given by $E(\chi'^2) = \phi + \lambda$ and $V(\chi'^2) = 2(\phi + 2\lambda)$, respectively.

Furthermore, let $\chi_1'^2$ and χ_2^2 be independent. Then the probability distribution of the following random variable F' is called the *noncentral F distribution* with (ϕ_1, ϕ_2) degrees of freedom and a noncentrality parameter λ (given by Eq. 1.58):

$$F' = \frac{\chi_1'^2/\phi_1}{\chi_2^2/\phi_2}, \tag{1.62}$$

and is denoted by $F'(\phi_1, \phi_2; \lambda)$. Since $\lambda = 0$ reduces $\chi'^2(\phi_1, \lambda)$ to $\chi^2(\phi_1)$, it also reduces $F'(\phi_1, \phi_2; \lambda)$ to $F(\phi_1, \phi_2)$. That is, the noncentral F distribution generalises the (central) F distribution.

For the sake of completeness, the pdf of $F'(\phi_1, \phi_2; \lambda)$ is provided below [13][22]:

$$f(x; \phi_1, \phi_2, \lambda) =$$

$$e^{-\frac{\lambda}{2}} g(x; \phi_1, \phi_2) \sum_{j=0}^{\infty} \left\{ \frac{\lambda \phi_1 x/(2\phi_2)}{1 + \phi_1 x/\phi_2} \right\}^j$$

$$\frac{(\phi_1 + \phi_2)(\phi_1 + \phi_2 + 2) \cdots \{\phi_1 + \phi_2 + 2(j-1)\}}{j! \phi_1(\phi_1 + 2) \cdots \{\phi_1 + 2(j-1)\}}, \tag{1.63}$$

where

$$g(x; \phi_1, \phi_2) = \frac{1}{B(\phi_1/2, \phi_2/2)} \left(\frac{\phi_1}{\phi_2}\right)^{\phi_1/2} x^{\phi_1/2-1} \left(1 + \frac{\phi_1 x}{\phi_2}\right)^{-(\phi_1+\phi_2)/2}. \tag{1.64}$$

Theorem 12 *If a random variable F' obeys $F'(\phi_1, \infty; \lambda)$, then $\phi_1 F'$ obeys $\chi'^2(\phi_1, \lambda)$* [13].

Corollary 15 *If a random variable F obeys $F(\phi_1, \infty)$, then $\phi_1 F$ obeys $\chi^2(\phi_1)$.*

Proof This is a special case of Theorem 12, where we let $\lambda = 0$.

Corollary 16 *Let F' and χ'^2 be random variables that obey $F'(\phi_A, \phi_E; \lambda)$ and $\chi'^2(\phi_A, \lambda)$, respectively. If ϕ_E can be approximated as $\phi_E = \infty$, then the following approximation holds:*

$$Pr\{F' \geq F_{inv}(\phi_A, \phi_E; P)\} \approx Pr\{\chi'^2 \geq \chi_{inv}^2(\phi_A; P)\}. \tag{1.65}$$

Proof Note that $Pr\{F' \geq F_{inv}(\phi_A, \infty; P)\} = Pr\{\phi_A F' \geq \phi_A F_{inv}(\phi_A, \infty; P)\}$. Now, from Theorem 12, $\phi_A F'$ may be rewritten as χ'^2, whereas, from Corollary 15,

---

[22]It is known that the population mean and the population variance of F' are given by $E(F') = \frac{\phi_2(\phi_1+\lambda)}{\phi_1(\phi_2-2)}$ (for $\phi_2 \geq 3$) and $V(F') = 2\left(\frac{\phi_2}{\phi_1}\right)^2 \frac{(\phi_1+\lambda)^2+(\phi_1+2\lambda)(\phi_2-2)}{(\phi_2-2)^2(\phi_2-4)}$ (for $\phi_2 \geq 5$), respectively.

$\phi_A F_{inv}(\phi_A, \infty; P)$ may be rewritten as $\chi^2_{inv}(\phi_A; P)$. Hence the above corollary holds.

Theorem 13 (Approximating a noncentral F distribution with a normal distribution) *For random variables F' and u that obey $F'(\phi_A, \phi_E; \lambda)$ and $N(0, 1^2)$, respectively, the following approximation holds* [13]:

$$Pr\{F' \le w\} \approx Pr \left\{ u \le \frac{\sqrt{\frac{w}{\phi_E}}\sqrt{2\phi_E - 1} - \sqrt{\frac{c_A}{\phi_A}}\sqrt{2\phi_A^* - 1}}{\sqrt{\frac{c_A}{\phi_A} + \frac{w}{\phi_E}}} \right\}, \qquad (1.66)$$

where

$$c_A = \frac{\phi_A + 2\lambda}{\phi_A + \lambda}, \quad \phi_A^* = \frac{(\phi_A + \lambda)^2}{\phi_A + 2\lambda}. \qquad (1.67)$$

A gist of the proof is given in Appendix 2 of Sakai [16].

References

1. C. Buckley, E.M. Voorhees, Retrieval system evaluation, in *TREC: Experiment and Evaluation in Information Retrieval*, Chap. 3, ed. by E.M. Voorhees, D.K. Harman (The MIT Press, Cambridge, 2005), pp. 53–75
2. B. Carterette, Bayesian inference for information retrieval evaluation, in *Proceedings of ACM ICTIR*, Northampton, 2015, pp. 31–40
3. J. Cohen, *Statistical Power Analysis for the Bahavioral Sciences*, 2nd edn. (Psychology Press, New York, 1988)
4. G.V. Cormack, C.R. Palmer, C.L.A. Clarke, Efficient construction of large test collections, in *Proceedings of ACM SIGIR*, Melbourne, 1998, pp. 282–289
5. G. Cumming, *Understanding the New Statistics: Effect Sizes, Confidence Intervals, and Meta-Analysis* (Routledge, New York/London, 2012)
6. P.D. Ellis, *The Essential Guide to Effect Sizes* (Cambridge University Press, Cambridge/New York, 2010)
7. P. Good, *Permutation, Parametric, and Bootstrap Tests of Hypothesis*, 3rd edn. (Springer, New York, 2005)
8. R.J. Grissom, J.J. Kim, *Effect Sizes for Research*, 2nd edn. (Routledge, New York, 2012)
9. K. Järvelin, J. Kekäläinen, Cumulated gain-based evaluation of IR techniques. ACM TOIS **20**(4), 422–446 (2002)
10. J.K. Kruschke, *Doing Bayesian Data Analysis*, 2nd edn. (Elsevier, Amsterdam, 2015)
11. K.R. Murphy, B. Myors, A. Wolach, *Statistical Power Analysis*, 4th edn. (Routledge, New York, 2014)
12. Y. Nagata, *How to Understand Statistical Methods (in Japanese)* (JUSE Press, Shibuya, 1996)
13. Y. Nagata, *How to Design the Sample Size (in Japanese)* (Asakura Shoten, Shinjuku, 2003)
14. S.E. Robertson, On document populations and measures of IR effectiveness, in *Proceedings of ICTIR*, Budapest, 2007, pp. 9–22
15. S.E. Robertson, E. Kanoulas, On per-topic variance in IR evaluation, in *Proceedings of ACM SIGIR*, Portland, 2012, pp. 891–900
16. T. Sakai, Topic set size design. Inf. Retr. **19**(3), 256–283 (2016)

17. T. Sakai, The probability that your hypothesis is correct, credible intervals, and effect sizes for IR evaluation, in *Proceedings of ACM SIGIR*, Shinjuku, 2017, pp. 25–34

18. H. Toyoda, (ed.), *Fundamentals of Bayesian Statistics: Practical Getting Started by Hamiltonian Monte Carlo Method (in Japanese)* (Asakura Shoten, Shinjuku, 2015)

19. H. Toyoda. *An Introduction to Statistical Data Analysis: Bayesian Statistics for 'post p-value era' (in Japanese)* (Asakuha Shoten, Shinjuku, 2016)

Chapter 2
t-Tests

Abstract This chapter first explains how the following classical significance tests for comparing two means work: the *paired t-test* for *paired data* (Sect. 2.2) and *(Student's) two-sample t-test* and *Welch's two-sample t-test* for *unpaired data* (Sects. 2.3 and 2.4). You have paired data if, for example, you evaluate two search engines using the same topic set with some evaluation measure such as *normalised Discounted Cumulative Gain* (nDCG) (Järvelin and Kekäläinen, ACM TOIS 20(4):422–446, 2002). (For a survey on IR evaluation measures, see Sakai (Metrics, statistics, tests. In: PROMISE winter school 2013: bridging between information retrieval and databases. LNCS 8173, pp 116–163, 2014).) You have unpaired data if, for example, you evaluated System 1 with User Group A and System 2 with User Group B; the group sizes may differ. This chapter then discusses the relationship between the aforementioned two two-sample *t*-tests (Sect. 2.5) and shows how the three *t*-tests can easily be conducted using Excel (Sect. 2.6) and R (Sect. 2.7). Finally, it describes how confidence intervals for the mean differences can be constructed, based on the assumptions that form the basis of the three *t*-tests (Sect. 2.8).

Keywords Confidence intervals · Paired *t*-test · Student's *t*-test · Two-sample *t*-test · Welch's *t*-test

2.1 Introduction

In what follows, I shall start with the assumption that each evaluation measure score for the same system independently obeys an identical normal distribution. However, the reader should note that the *t*-test is in fact highly robust to the normality assumption violation. Recall the Central Limit Theorem from Chap. 1 Sect. 1.2.2: the theorem says that the sample mean of identically distributed values such as nDCG scores approximately obeys a normal distribution *even if we do not know what the original identical distribution is, for sufficiently large samples*. Hence, while we utilise Corollaries 5 and 6 from Chap. 1 Sect. 1.2.2 (for paired and unpaired data, respectively) to obtain random variables that should obey a *t*-distribution

© Springer Nature Singapore Pte Ltd. 2018

T. Sakai, *Laboratory Experiments in Information Retrieval*,
The Information Retrieval Series 40, https://doi.org/10.1007/978-981-13-1199-4_2

under the null hypothesis, we can *relax* the prerequisites of the corollaries, namely, that the original identical distributions must be normal. For example, given a large sample size n, Corollary 5 can be interpreted more loosely as[1]: "If $x_1, x_2, \ldots, x_n$ are independent and identically distributed such that $E(x_j) = \mu$ and $V(x_j) = \sigma^2$, then $t = (\bar{x} - \mu)/\sqrt{V/n}$ approximately obeys $t(n - 1)$".

Despite the above time-honoured wisdom, the IR community appears to have been historically rather reluctant to use parametric tests such as the t-test [9]. Salton and Lesk, 1968 [11] (p. 15): "since this normality is generally hard to prove for statistics derived from a request-document correlation process, the sign test probabilities may provide a better indicator of system performance than the t-test"; Van Rijsbergen, 1979 [18] (p. 136): "Parametric tests are inappropriate because we do not know the form of the underlying distribution"; Sparck Jones and Willet, 1997 [14] (p. 170): "Since the form of the population distributions underlying the observed performance values is not known, only weak tests can be applied; for example, the sign test." In contrast, Tague-Sutcliffe's remark from 1992 [16] indirectly refers to the Central Limit Theorem (Chap. 1 Sect. 1.2.2): "Frequently, whether or not a statistic has a normal or related distribution is dependent on the sample size, with statistics from small samples being less likely to be normally distributed. Thus, nonparametric methods are most often used with small samples." Moreover, Hull's SIGIR 1993 paper [2] may have encouraged some IR researchers to take up parametric tests: he argued that the normality assumption may be tested using diagnostic plots and described the *paired t-test* (covered in Sect. 2.2 in this chapter), *two-way ANOVA without replication* (covered in Chap. 3 Sect. 3.2), *Tukey's honestly significant difference test with paired observations* (covered in Chap. 4 Sect. 4.4.3), along with their nonparametric alternatives.

The advent of high-performance computers have enabled us to conduct computer-based significance tests such as the *bootstrap* and *randomisation tests* (See Chap. 4 Sect. 4.5.1) as alternatives to classical tests such as the t-test. These computer-based tests are completely free from the normality assumption yet yield p-values that are similar to those from the t-test. The robustness of the t-test can thus be demonstrated empirically also. Savoy in 1997 [12] and Sakai in 2006 [7] advocated the use of the bootstrap test; in 2007, Smucker, Allan, and Carterette [13] used TREC ad hoc runs to compare the paired t-test, bootstrap, and randomisation tests, along with the nonparametric Wilcoxon signed-rank and sign tests, and concluded that the first three tests yield comparable p-values and went so far as to say that "IR researchers should discontinue use of the Wilcoxon and sign tests."[2]

---

[1] See Chap. 1 Sect. 1.2.2 for a brief discussion on: How large is "sufficiently large?".

[2] In contrast, in 1998, Zobel [21] conducted topic set splitting experiments with early TREC data to compare parametric and nonparametric tests and recommended Wilcoxon signed-rank test over the paired t-test and ANOVA. Moreover, in 2013, Urbano, Marrero, and Martín [17] reported that the Wilcoxon test, the paired t-test, and the bootstrap are more reliable than the randomisation test.

2.2 Paired *t*-Test

You evaluated two systems using a test collection with n topics. You have a mean nDCG score averaged over the n topics for System 1 and another for System 2; you want to discuss whether the observed difference between the two systems is "real".

Let x_{1j} denote the nDCG score of System 1 for the j-th topic; similarly, let x_{2j} denote the nDCG score of System 2 for the j-th topic ($j = 1, \ldots, n$). We regard them as random variables, and the actual scores we have on the table as realisations of these variables. Let us assume that x_{1j} and x_{2j} are all independent of one another and that:

$$x_{1j} \sim N(\mu_1, \sigma_1^2), \quad x_{2j} \sim N(\mu_2, \sigma_2^2), \tag{2.1}$$

where "$\sim$" means "(a random variable) obeys (a probability distribution)".

Consider per-topic score differences $d_j = x_{1j} - x_{2j}$ ($j = 1, \ldots, n$). Under the above normality assumptions, Theorem 4 (Chap. 1 Sect. 1.2.2) gives us:

$$d_j \sim N(\mu_1 - \mu_2, \sigma_1^2 + \sigma_2^2). \tag{2.2}$$

Moreover, under the above independence assumptions, d_j's are also independent. Therefore, from Corollary 5 (Chap. 1 Sect. 1.2.4), we know that

$$t = \frac{\bar{d} - (\mu_1 - \mu_2)}{\sqrt{V_d/n}} \sim t(n-1), \tag{2.3}$$

where

$$\bar{d} = \frac{1}{n} \sum_{j=1}^{n} d_j, \tag{2.4}$$

and

$$V_d = \frac{\sum_{j=1}^{n}(d_j - \bar{d})^2}{n-1} = \frac{\sum_{j=1}^{n} d_j^2 - (\sum_{j=1}^{n} d_j)^2/n}{n-1}. \tag{2.5}$$

Note that, from Theorem 6 (Chap. 1 Sect. 1.2.2), V_d is an unbiased estimator of the population variance $\sigma_1^2 + \sigma_2^2$.

Given the observed nDCG scores, are the population means of the two systems different? To discuss this, we let the null hypothesis be $H_0 : \mu_1 = \mu_2$ and the alternative hypothesis be $H_1 : \mu_1 \neq \mu_2$ (a two-sided test). Given Eq. 2.3, if we further assume H_0, we obtain:

$$t_0 = \frac{\bar{d}}{\sqrt{V_d/n}} \sim t(n-1). \tag{2.6}$$

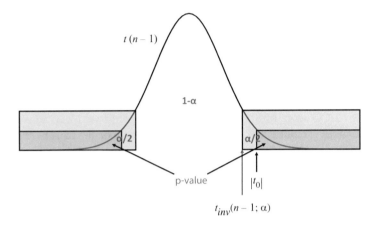

$t(n-1)$

$1-\alpha$

$\alpha/2$ $\alpha/2$

p-value $|t_0|$

$t_{inv}(n-1;\alpha)$

Fig. 2.1 A situation where H_0 is rejected since $|t_0| \geq t_{inv}(n-1;\alpha)$

Therefore, for a given significance criterion α, if $|t_0| \geq t_{inv}(n-1;\alpha)$, we reject H_0 because such an extreme t_0 value is highly unlikely to occur if it really obeys $t(n-1)$; we say that the difference between the two systems is statistically significant at α. Figure 2.1 depicts this situation. On the other hand, if $|t_0| < t_{inv}(n-1;\alpha)$, we say that the difference is not statistically significant.

2.3   Two-Sample *t*-Test

One system was evaluated using a set of n_1 topics; another was evaluated using a different set, containing n_2 topics. Thus you have a mean nDCG over n_1 topics for System 1 and another over n_2 topics for System 2. You want to discuss whether the observed difference between the two systems is "real". Note that now we are dealing with unpaired data: each topic from the first set generally does not correspond to any from the second set. Alternatively, we can consider a situation where one system was evaluated by User Group A and another with User Group B, and user ratings were obtained for each system. However, henceforth we speak of topic sets rather than user groups.

Scores of System 1 are denoted by x_{1j} ($j = 1, \ldots, n_1$), and scores of System 2 are denoted by x_{2j} ($j = 1, \ldots, n_2$). As with the paired *t*-test, we assume that x_{1j} and x_{2j} are all independent of one another and that:

$$x_{1j} \sim N(\mu_1, \sigma^2), \quad x_{2j} \sim N(\mu_2, \sigma^2). \tag{2.7}$$

Note the difference between this and Eq. 2.1: we are now making a stronger assumption, which says that the two systems share a common variance, σ^2. This is called the *homoscedasticity* assumption. While homoscedasticity is a strong assumption, the t-test is in fact quite robust to violations of this assumption, as we shall discuss in Sect. 2.5.

From Eq. 2.7 and Corollary 6 (Chap. 1, Sect. 1.2.4), we know that

$$t = \frac{\bar{x}_{1\bullet} - \bar{x}_{2\bullet} - (\mu_1 - \mu_2)}{\sqrt{V_p(1/n_1 + 1/n_2)}} \sim t(n_1 + n_2 - 2) , \tag{2.8}$$

where, as we have discussed in Chap. 1 Sect. 1.2.4,

$$\bar{x}_{1\bullet} = \frac{\sum_{j=1}^{n_1} x_{1j}}{n_1} , \qquad \bar{x}_{2\bullet} = \frac{\sum_{j=1}^{n_2} x_{2j}}{n_2} , \tag{2.9}$$

and

$$S_1 = \sum_{j=1}^{n_1}(x_{1j} - \bar{x}_{1\bullet})^2 , \qquad S_2 = \sum_{j=1}^{n_2}(x_{2j} - \bar{x}_{2\bullet})^2 , \qquad V_p = \frac{S_1 + S_2}{n_1 + n_2 - 2} . \tag{2.10}$$

As we did with the paired t-test, let us consider a two-sided test, with H_0, $\mu_1 = \mu_2$, and H_1, $\mu_1 \neq \mu_2$. Given Eq. 2.8, if we further assume H_0, we obtain:

$$t_0 = \frac{\bar{x}_{1\bullet} - \bar{x}_{2\bullet}}{\sqrt{V_p(1/n_1 + 1/n_2)}} \sim t(n_1 + n_2 - 2) . \tag{2.11}$$

Therefore, for a given significance criterion α, if $|t_0| \geq t_{inv}(n_1 + n_2 - 2; \alpha)$, we reject H_0; we say that the difference between the two systems is statistically significant at α. On the other hand, if $|t_0| < t_{inv}(n_1 + n_2 - 2; \alpha)$, we say that the difference is not statistically significant.

2.4 Welch's Two-Sample t-Test

There is an alternative version of the two-sample t-test, known as *Welch's t-test*, which adheres to the model given by Eq. 2.1 rather than Eq. 2.7. That is, Welch's t-test does not rely on the homoscedasticity assumption. To avoid confusion, henceforth we shall refer to the two-sample t-test described in Sect. 2.3 as *Student's t-test*.[3] The reader should be made aware at this point, however, that Welch's t-test

---

[3] In 1908, William Sealy Gosset, who worked for Arthur Guinness Son & Co., Ltd., published his seminal paper on the t distribution under the pseudonym "Student" [15, 19]. There is no mention of "t" in Gosset's original paper [15]; the test statistic is referred to as "z" there. "In 1912, Fisher, while still an undergraduate at Cambridge, made a tiny correction to Gosset's z, and in 1922 they agreed to rename the corrected tables and test "Student's" t" [20].

should *not* be regarded as something superior to Student's t-test: unlike Student's t-test, Welch's t-test involves *approximations*.

The procedure for Welch's t-test is identical to that for Student's, except for the computation of the test statistic and the degrees of freedom (Eq. 2.11). For Welch's t-test, the test statistic is given by:

$$t_{w0} = \frac{\bar{x}_{1\bullet} - \bar{x}_{2\bullet}}{\sqrt{V_1/n_1 + V_2/n_2}} \; , \tag{2.12}$$

where the two sample variances V_1 and V_2 are defined as in Eq. 1.35. As for the degrees of freedom for the t-distribution, it is computed as:

$$\phi^* = \left(\frac{V_1}{n_1} + \frac{V_2}{n_2} \right)^2 / \left\{ \frac{(V_1/n_1)^2}{n_1 - 1} + \frac{(V_2/n_2)^2}{n_2 - 1} \right\} . \tag{2.13}$$

Thus, iff $|t_{w0}| \geq t_{inv}(\phi^*; \alpha)$, we reject H_0.

Welch's t-test *approximates* the distribution of the following random variable by a χ^2 distribution with ϕ_0 degrees of freedom [6]:

$$W = \phi_0 \left(\frac{V_1}{n_1} + \frac{V_2}{n_2} \right) / \left(\frac{\sigma_1^2}{n_1} + \frac{\sigma_2^2}{n_2} \right) . \tag{2.14}$$

Furthermore, it estimates ϕ_0 as $\hat{\phi}_0 = \phi^*$, using Eq. 2.13. For these two reasons, choosing Welch's t-test over Student's *unconditionally* is not recommended. Given a two-sample situation, which t-test should we choose, then? This is the topic of the next section.

2.5   Which Two-Sample t-Test?

The bottom line first: use Student's t-test *unless* the two sample sizes n_1 and n_2 are *very* different (e.g., one is twice as large as the other).

Nagata [6] clarifies the relationship between the two two-sample t-tests. First, let us define the *sample size ratio a* and the *sample variance ratio b* as follows:

$$a = \frac{n_2}{n_1} \; , \quad b = \frac{V_2}{V_1} \; . \tag{2.15}$$

From Eqs. 2.11 and 2.12, the ratio of the two test statistics can be derived as:

$$G(a, b) = \frac{t_{w0}}{t_0} = \sqrt{\frac{V_p(\frac{1}{n_1} + \frac{1}{n_2})}{\frac{V_1}{n_1} + \frac{V_2}{n_2}}} = \sqrt{\frac{(a + 1)\{n_1(ab + 1) - (b + 1)\}}{(a + b)\{n_1(a + 1) - 2\}}} .$$

$$(2.16)$$

Note that $G(1, b) = G(a, 1) = 1$. Hence, if either $n_1 = n_2$ or $V_1 = V_2$ holds, then $t_0 = t_{w0}$ holds: the two test statistics are equal. In practice, if the larger sample is no larger than 1.5 times the other *or* if the larger variance is no larger than 1.5 times the other, t_0 and t_{w0} will differ by at most 20% or so [6].

As for the degrees of freedom for the t-distribution, recall that while Student's t-test uses $\phi = n_1 + n_2 - 2$, Welch's t-test uses ϕ^* given by Eq. 2.13. It can be shown that [6]:

$$H(a, b) = \frac{\phi^*}{\phi} = \frac{(a + b)^2(n_1 - 1)(an_1 - 1)}{\{a^2(an_1 - 1) + b^2(n_1 - 1)\}\{(a + 1)n_1 - 2\}} .$$

$$(2.17)$$

It can be verified that $H(1, 1) = 1$ holds. Hence, having $n_1 = n_2$ and $V_1 = V_2$ is a sufficient condition for obtaining $\phi^* = \phi$. Furthermore, it can be verified that $H(a, b) = \phi^*/\phi$ is much smaller than one if $b = V_2/V_1$ is close to one *and a =* n_2/n_1 is far from one. That is, Welch's t-test has a relatively low statistical power when the variances are roughly equal but the sample sizes are quite different, since a small ϕ^* implies high uncertainty, i.e. fatter tails for the t-distribution.

In Sakai [10], I compared Student's and Welch's t-test in the context of IR evaluation. The results suggested that if the sample sizes differ substantially *and* if the larger sample has a substantially larger variance, Welch's t-test may be less reliable. In practice, note that, unlike the variances, the samples sizes are relatively easy for the researchers to control. Hence, in IR evaluation, I recommend Student's t-test as the first choice, with a prerequisite that similar sample sizes are used wherever possible.

Some textbooks recommend conducting an F-test for testing whether $\sigma_1^2 = \sigma_2^2$ first and then deciding whether to use Student's t-test or Welch's t-test based on the outcome of the F-test. Nagata [5] recommends that if the F-test must be conducted for the above purpose, the Type I error rate α for that test should be set to 20% rather than 5%, so that the Type II error rate β (i.e. the probability of missing the fact that $\sigma_1^2 \neq \sigma_2^2$) is kept low. However, as we have discussed above, Student's t-test is reliable even for unequal variances provided that the two sample sizes are similar. Hence the F-test for checking homoscedasticity seems unnecessary at least for similar sample sizes.

2.6   Conducting a t-Test with Excel

Now that we know the assumptions behind the t-tests and the procedures for conducting them, let us actually conduct t-tests using real data. The reader can download a sample *topic-by-run matrix* in csv format from http://www.f.waseda.

A23	▼	⋮	✕	✓	*fx*	=T.TEST(A2:A21, B2:B21, 2, 1)	

	A	B	C	D	E	F	G
1	System1	System2	System3				
2	0.4695	0.3732	0.3575				
3	0.2813	0.3783	0.2435				
4	0.3914	0.3868	0.3167				
5	0.6884	0.5896	0.6024				
6	0.6121	0.4725	0.4766				
7	0.3266	0.233	0.2429				
8	0.5605	0.4328	0.4066				
9	0.5916	0.5073	0.4707				
10	0.4385	0.3889	0.3384				
11	0.5821	0.5551	0.4597				
12	0.2871	0.3274	0.2769				
13	0.5186	0.5066	0.4066				
14	0.5188	0.5198	0.3859				
15	0.5019	0.4981	0.4568				
16	0.4702	0.3878	0.3437				
17	0.329	0.4387	0.2649				
18	0.4758	0.4946	0.4045				
19	0.3028	0.34	0.3253				
20	0.3752	0.4895	0.3205				
21	0.2796	0.2335	0.224				
22							
23	0.2057765						

Fig. 2.2 The "laziest" way to obtain a *p*-value with a paired *t*-test using Microsoft Excel

jp/tetsuya/20topics3runs.mat.csv: the file has 21 rows (representing 20 topics plus a header row) and 3 columns (representing 3 runs); each cell is an evaluation measure score. The csv file can be opened using Microsoft Excel.[4] Since a *t*-test concerns a comparison of two means, let us use the first two columns (i.e. runs) of this file in this chapter.

Figure 2.2 shows the "laziest" way to obtain a *p*-value with a paired *t*-test for comparing the first two systems: it can be seen that, by just entering T.TEST(A2:A21, B2:B21, 2, 1), the *p*-value (about 0.2058) is obtained. The third argument 2 means that we want a two-sided test,[5] and the fourth argument 1 means that we want a *paired t*-test. Try changing the fourth argument to 2 and to 3: this will give you the *p*-values based on Student's and Welch's two-sample *t*-tests, respectively. That is, the 20 values in Column 1 and the 20 values in Column 2 are treated as unpaired data. It can be verified that the *p*-values with these two tests

---

[4]At the time of this writing, I am using Microsoft Office 2013, but later versions will probably support all the functionalities discussed in this book.

[5]Set the third argument to 1 if a one-sided test is needed.

	A	B	C	D	E	F	G
E21			f_x	=A21−B21			
	System1	System2	System3		S1−S2		
1	System1	System2	System3		S1−S2		
2	0.4695	0.3732	0.3575		0.0963		
3	0.2813	0.3783	0.2435		−0.097		
4	0.3914	0.3868	0.3167		0.0046		
5	0.6884	0.5896	0.6024		0.0988		
6	0.6121	0.4725	0.4766		0.1396		
7	0.3266	0.233	0.2429		0.0936		
8	0.5605	0.4328	0.4066		0.1277		
9	0.5916	0.5073	0.4707		0.0843		
10	0.4385	0.3889	0.3384		0.0496		
11	0.5821	0.5551	0.4597		0.027		
12	0.2871	0.3274	0.2769		−0.0403		
13	0.5186	0.5066	0.4066		0.012		
14	0.5188	0.5198	0.3859		−0.001		
15	0.5019	0.4981	0.4568		0.0038		
16	0.4702	0.3878	0.3437		0.0824		
17	0.329	0.4387	0.2649		−0.1097		
18	0.4758	0.4946	0.4045		−0.0188		
19	0.3028	0.34	0.3253		−0.0372		
20	0.3752	0.4895	0.3205		−0.1143		
21	0.2796	0.2335	0.224		0.0461		
22							
23	0.2057765			dbar	0.022375		
24				Vt	0.0058336		
25				t0	1.310113		
26				p−value	0.2057765		

Fig. 2.3 The step-by-step method for obtaining a p-value with a paired t-test using Microsoft Excel

are much larger than that with the paired t-test and that these two p-values are very similar (about 0.5300 and 0.5302, respectively). Thus, a paired test should always be used if the data are really paired. Furthermore, all three tests suggest that the two systems are not statistically significantly different from each other.

When using Excel, however, I do *not* recommend the use of the T.TEST function, because it makes one treat the t-test as a black box. For example, one cannot discuss the *effect size* (See Chap. 5) if she just relies on this function. Moreover, for Welch's t-test, the T.TEST function does not even tell us what the (estimated) degrees of freedom ϕ^* is. In research papers, such details should always be reported together with the significance test results, as we shall discuss in Chap. 5.

Given a csv file like the one shown in Fig. 2.2, even a step-by-step approach to conducting the t-test is very easy. Figure 2.3 shows an example for conducting a paired t-test. First, enter "=A2-B2" in Cell E2, and drag it down to Cell E21, so that d_j is obtained for $j = 1, \ldots, 20$ (See Eq. 2.2). Then:

$\bar{d}$ = AVERAGE(E2:E21) = 0.022375   (Eq. 2.4);
V_d = DEVSQ(E2:E21)/(20 − 1) = 0.005834   (Eq. 2.5);
t_0 = $\bar{d}/\sqrt{V_d/20}$ = 1.3101   (Eq. 2.6);
p-value = T.DIST.2T($|t_0|$, 20 − 1) = 0.2058 .

Here, T.DIST.2T($|t_0|$, ϕ) returns the two-tailed probability of observing $|t_0|$ or something more extreme when t_0 obeys a *t*-distribution with ϕ degrees of freedom (i.e. the two purple tails in Fig. 2.1). Figure 2.3 shows that the *p*-value thus obtained is the same as the one obtained using T.TEST.

Similarly, using the sample data shown in Fig. 2.3, Student's *t*-test can be conducted step-by-step as follows:

$\bar{x}_{1\bullet} − \bar{x}_{2\bullet}$ = AVERAGE(A2:A21) − AVERAGE(B2:B21) = 0.022375;
V_p = (DEVSQ(A2:A21) + DEVSQ(B2:B21))/(20 + 20 − 2)
        = 0.012463   (Eq. 1.29);
t_0 = $(\bar{x}_{1\bullet} − \bar{x}_{2\bullet})/\sqrt{V_p(1/20 + 1/20)}$ = 0.6338   (Eq. 2.11);
p-value = T.DIST.2T($|t_0|$, 20 + 20 − 2) = 0.5300.

On the other hand, Welch's *t*-test can be conducted as follows:

V_1 = DEVSQ(A2:A21)/(20 − 1) = 0.015323   (Eq. 1.35);
V_2 = DEVSQ(B2:B21)/(20 − 1) = 0.009602;
t_{w0} = $(\bar{x}_{1\bullet} − \bar{x}_{2\bullet})/\sqrt{V_1/20 + V_2/20}$ = 0.6338   (Eq. 2.12);
ϕ^* = 36.0985   (Eq. 2.13);
p-value = T.DIST.2T($|t_{w0}|$, ϕ^*) = 0.5302.

2.7   Conducting a *t*-Test with R

Now let us use the data we used in Sect. 2.6 and conduct the same *t*-tests, using R instead of Excel this time. R can easily be downloaded and installed from http://cran.r-project.org/: for details, please refer to textbooks on R (e.g. [1, 4]).

First, let us load the same topic-by-run matrix onto R from the aforementioned URL:

```
> matURL <-
+ "http://www.f.waseda.jp/tetsuya/20topics3runs.
   mat.csv"
> (mat <-
+ read.table(file = matURL, header = TRUE, sep = ","))
```

Note that ">" and "+" are prompts from the R command line interface and not part of the commands entered. The above two commands store the topic-by-run matrix into the variable mat. Figure 2.4 verifies that the scores are stored correctly: note that the second command is enclosed within " (" and ") " just to display the contents of mat.

```
> matURL <-
+ "http://www.f.waseda.jp/tetsuya/20topics3runs.mat.csv"
> (mat <-
+ read.table(file = matURL, header = TRUE, sep = ","))
    System1 System2 System3
1    0.4695  0.3732  0.3575
2    0.2813  0.3783  0.2435
3    0.3914  0.3868  0.3167
4    0.6884  0.5896  0.6024
5    0.6121  0.4725  0.4766
6    0.3266  0.2330  0.2429
7    0.5605  0.4328  0.4066
8    0.5916  0.5073  0.4707
9    0.4385  0.3889  0.3384
10   0.5821  0.5551  0.4597
11   0.2871  0.3274  0.2769
12   0.5186  0.5066  0.4066
13   0.5188  0.5198  0.3859
14   0.5019  0.4981  0.4568
15   0.4702  0.3878  0.3437
16   0.3290  0.4387  0.2649
17   0.4758  0.4946  0.4045
18   0.3028  0.3400  0.3253
19   0.3752  0.4895  0.3205
20   0.2796  0.2335  0.2240
```

Fig. 2.4 Loading the topic-by-run matrix with R

As before, let us conduct *t*-tests by focussing on the first two columns. The following commands execute the paired *t*-test, Student's two-sample *t*-test, and Welch's two-sample *t*-test, respectively:

```
> t.test(mat$System1, mat$System2, paired = TRUE)
> t.test(mat$System1, mat$System2, var.equal = TRUE)
> t.test(mat$System1, mat$System2, var.equal = FALSE)
```

Figures 2.5 and 2.6 show the results displayed. The reader should check that the *t*-values and *p*-values are consistent with the Excel results given in Sect. 2.6. Note that Fig. 2.6 includes the result when the t.test function is used with two arguments only: it can be observed that Welch's *t*-test was called as the default.

```
> t.test(mat$System1, mat$System2)
```

The significance test results with R contain not only the *p*-value but also the *t*-value t_0 (shown in the output as t) and the degrees of freedom ϕ (shown in the output as df). They also contain 95% *confidence intervals*, which we shall discuss this in the next section.

```
> t.test(mat$System1, mat$System2, paired = TRUE)

        Paired t-test

data:  mat$System1 and mat$System2
t = 1.3101, df = 19, p-value = 0.2058
alternative hypothesis: true difference in means is not equal to 0
95 percent confidence interval:
 -0.01337109  0.05812109
sample estimates:
mean of the differences
              0.022375


> t.test(mat$System1, mat$System2, var.equal = TRUE)

        Two Sample t-test

data:  mat$System1 and mat$System2
t = 0.6338, df = 38, p-value = 0.53
alternative hypothesis: true difference in means is not equal to 0
95 percent confidence interval:
 -0.04909139  0.09384139
sample estimates:
mean of x mean of y
 0.450050  0.427675
```

Fig. 2.5 Conducting a paired *t*-test and a Student's *t*-test with R

```
> t.test(mat$System1, mat$System2, var.equal = FALSE)

        Welch Two Sample t-test

data:  mat$System1 and mat$System2
t = 0.6338, df = 36.098, p-value = 0.5302
alternative hypothesis: true difference in means is not equal to 0
95 percent confidence interval:
 -0.04921523  0.09396523
sample estimates:
mean of x mean of y
 0.450050  0.427675


> t.test(mat$System1, mat$System2)

        Welch Two Sample t-test

data:  mat$System1 and mat$System2
t = 0.6338, df = 36.098, p-value = 0.5302
alternative hypothesis: true difference in means is not equal to 0
95 percent confidence interval:
 -0.04921523  0.09396523
sample estimates:
mean of x mean of y
 0.450050  0.427675
```

Fig. 2.6 Conducting a Welch's *t*-test with R

2.8   Confidence Intervals for the Difference in Population Means

In the discussion of the paired t-test, we obtained Eq. 2.3, which we repeat here:

$$t = \frac{\bar{d} - (\mu_1 - \mu_2)}{\sqrt{V_d/n}} \sim t(n-1) . \tag{2.18}$$

Since t obeys $t(n-1)$, for a given Type I probability α, we have:

$$Pr\{-t_{inv}(n-1;\alpha) \le t \le t_{inv}(n-1;\alpha)\} = 1 - \alpha . \tag{2.19}$$

From the above, we obtain:

$$Pr\{\bar{d} - MOE \le \mu_1 - \mu_2 \le \bar{d} + MOE\} = 1 - \alpha , \tag{2.20}$$

where the *margin of error* (MOE) is given by:

$$MOE = t_{inv}(n-1;\alpha)\sqrt{V_d/n} . \tag{2.21}$$

Equation 2.20 implies that the $100(1 - \alpha)\%$ *confidence interval* (CI) for the difference between the two population means is given by $[\bar{d} - MOE, \bar{d} + MOE]$.

Let $\alpha = 0.05$ to discuss 95% CIs. What does a 95% CI mean? An explanation we occasionally come across is: "It means that the population difference will fall into the above range with 95% probability". However, this interpretation is rather misleading. This is because, as we have discussed earlier, we view population parameters as constants, and therefore the difference in population means ($\mu_1 - \mu_2$) is also a constant, not a random variable. The random variables are those associated with the actual *samples*, not the populations. Thus, a more accurate explanation for a 95% CI would be: "We take a sample, and construct a CI; we take another sample, and construct another CI. we do this (say) 100 times in total. Then, about 95 of the 100 different CIs will actually capture the true population difference".

Let us use Excel to compute the 95% CI for the difference in population means using the paired data discussed in Sect. 2.6, with $\bar{d} = 0.022375$, $V_d = 0.008534$:

$t_{inv}(n-1; 0.05)$ $= \texttt{T.INV.2T}(0.05, 20 - 1) = 2.093024$;
MOE $= 0.035746$ (Eq. 2.21);
95%CI $= [-0.0134, 0.0581]$.

It can be observed that the result is consistent with what R gave us in Fig. 2.5 ("`Paired t-test`"). Again, the 95% CI is "probably right" in the sense that about 95% of different CIs obtained from different samples will actually contain the true difference in population means. However, in this example, we cannot say much about which population mean may actually be larger, since the CI for the difference in means covers both negative and positive values.

Let us obtain 95% CIs of the difference in population means for *two-sample* cases, based on the assumptions behind Student's t-test as well as Welch's t-test. From Student's t-test, recall Eq. 2.8, which we repeat here:

$$t = \frac{\bar{x}_{1\bullet} - \bar{x}_{2\bullet} - (\mu_1 - \mu_2)}{\sqrt{V_p(1/n_1 + 1/n_2)}} \sim t(n_1 + n_2 - 2) .$$

By using the same logic as the one used to obtain the 95% CI for the paired case, we have:

$$Pr\{\bar{x}_{1\bullet} - \bar{x}_{2\bullet} - MOE \le \mu_1 - \mu_2 \le \bar{x}_{1\bullet} - \bar{x}_{2\bullet} + MOE\} = 1 - \alpha , \qquad (2.22)$$

where

$$MOE = t_{inv}(n_1 + n_2 - 2; \alpha)\sqrt{V_p(1/n_1 + 1/n_2)} . \qquad (2.23)$$

On the other hand, based on Welch's approximations, we can replace the above with the following MOE:

$$MOE = t_{inv}(\phi^*; \alpha)\sqrt{V_1/n_1 + V_2/n_2} , \qquad (2.24)$$

with ϕ^* computed using Eq. 2.13, and V_1, V_2 computed using Eq. 1.35. In either case, the $100(1 - \alpha)\%$ CI for the difference between the two population means is given by $[\bar{x}_{1\bullet} - \bar{x}_{2\bullet} - MOE, \ \bar{x}_{1\bullet} - \bar{x}_{2\bullet} + MOE]$.

Going back to our example, when the two columns are treated as unpaired data ($\bar{x}_{1\bullet} - \bar{x}_{2\bullet} = 0.022375$, $V_p = 0.012463$, $V_1 = 0.015323$, $V_2 = 0.009602$, $\phi^* = 36.0985$ from Sect. 2.6), based on Student's assumptions, we have:

$t_{inv}(n_1 + n_2 - 2; 0.05)$ $= \text{T.INV.2T}(0.05; 20 + 20 - 2) = 2.024394;$
MOE $= 0.071466$ (Eq. 2.23);
$95\%\text{CI}$ $= [-0.0491, 0.0938].$

On the other hand, based on Welch's assumptions and approximations, we have:

$t_{inv}(\phi^*; 0.05)$ $= \text{T.INV.2T}(0.05, 36.0985) = 2.028094;$
MOE $= 0.071597$ (Eq. 2.24);
$95\%\text{CI}$ $= [-0.0492, 0.0940].$

Compare these results with the ones R gave us in Fig. 2.5 ("two-sample t-test") and Fig. 2.6 ("Welch's two-sample t-test"). Note also that the two-sample CIs are substantially wider than the paired version and that the two two-sample CIs are very similar. Again, use the paired version if the data are really paired.

The above methods for constructing CIs for the difference between two population means are often used even if there are more than two systems. However, it should be noted that if multiple 95% CIs are constructed independently, this merely means that a CI for System i may capture its population mean difference about 95%

of the time, while a CI for System i' may capture its population mean difference about 95% of the time; it does *not* mean that these CIs capture their respective population mean differences *simultaneously*. *Simultaneous confidence intervals* for the difference between each system pair will be discussed in Chap. 4 Sect. 4.4.4.

References

1. M.J. Crawley, *Statistics: An Introduction Using R*, 2nd edn. (Wiley, Chichester, 2015)
2. D. Hull, Using statistical testing in the evaluation of retrieval experiments, in *Proceedings of ACM SIGIR'93*, Pittsburgh, 1993, pp. 329–338
3. K. Järvelin, J. Kekäläinen, Cumulated gain-based evaluation of IR techniques. ACM TOIS **20**(4), 422–446 (2002)
4. J.P. Lander, *R for Everyone* (Addison Wesley, Upper Saddle River, 2014)
5. Y. Nagata, *Introduction to Statistical Analysis (in Japanese)* (JUSE Press, Shibuya, 1992)
6. Y. Nagata, *How to Understand Statistical Methods (in Japanese)* (JUSE Press, Shibuya, 1996)
7. T. Sakai, Evaluating evaluation metrics based on the bootstrap, in *Proceedings of ACM SIGIR*, Seattle, 2006, pp. 525–532
8. T. Sakai, Metrics, statistics, tests, in *PROMISE Winter School 2013: Bridging Between Information Retrieval and Databases*, Bressanone. LNCS 8173, 2014, pp. 116–163
9. T. Sakai, Statistical reform in information retrieval? SIGIR Forum **48**(1), 3–12 (2014)
10. T. Sakai, Two-sample t-tests for IR evaluation: student or welch? in *Proceedings of ACM SIGIR*, Pisa, 2016, pp. 1045–1048
11. G. Salton, M.E. Lesk, Computer evaluation of indexing and text processing. J. ACM **15**(1), 8–36 (1968)
12. J. Savoy, Statistical inference in retrieval effectiveness evaluation. Inf. Process. Manag. **33**(4), 495–512 (1997)
13. M.D. Smucker, J. Allan, B. Carterette, A comparison of statistical significance tests for information retrieval evaluation, in *Proceedings of ACM CIKM, Lisbon*, 2007, pp. 623–632
14. K. Sparck Jones, P. Willet (eds.), *Readings in Information Retrieval* (Morgan Kaufmann, San Francisco, 1997)
15. Student, The probable error of a mean. Biometrika **6**, 1–25 (1908)
16. J. Tague-Sutcliffe, The pragmatics of information retrieval experimentation, revisited. Inf. Process. Manag. **28**, 467–490 (1992)
17. J. Urbano, M. Marrero, D. Martín, A comparison of the optimality of statistical significance tests for information retrieval evaluation, in *Proceedings of ACM SIGIR*, Dublin, 2013, pp. 925–928
18. C.J. van Rijsbergen, *Information Retrieval*, Chap. 7 (Butterworths, London, 1979)
19. S.L. Zabell, On student's 1908 article "the probable error of a mean". J. Am. Stat. Assoc. **103**(481), 1–7 (2008)
20. S.T. Ziliak, D.N. McCloskey, *The Cult of Statistical Significance: How the Standard Error Costs Us Jobs, Justice, and Lives* (The University of Michigan Press, Ann Arbor, 2008)
21. J. Zobel, How reliable are the results of large-scale information retrieval experiments? in *Proceedings of ACM SIGIR*, Melbourne, 1998, pp. 307–314

Chapter 3
Analysis of Variance

Abstract This chapter first describes the following classical *analysis of variance* (ANOVA) tests for comparing more than two means: *one-way ANOVA*, a generalised form of the unpaired t-test (Sect. 3.1); *two-way ANOVA without replication*, a generalised form of the paired t-test (Sect. 3.2); and *two-way ANOVA with replication*, which considers the interaction between two factors (e.g. topic and system). The first two types of ANOVA are particularly important for IR researchers, since, in laboratory experiments where systems are evaluated using topics, there is usually one evaluation measure score for a given topic-system pair, (unless, for example, the system is considered to be nondeterministic and produces a different search result page every time the same query is entered) where it is not possible to discuss the topic-system interaction. (Banks et al. (Inf Retr 1:7–34, 1999) applied *Tukey's single-degree-of-freedom test for nonadditivity* and *Mandel's bundle-of-line approach* to discuss topic-system interaction given two-way ANOVA *without* replication data from TREC-3 and reported: "there is a strong interaction between system and topic in terms of average precision. The presence of interaction implies that one cannot find simple descriptions of the data in terms of topics and systems alone." These tests are beyond the scope of this book.) This chapter then describes how one-way ANOVA and two-way ANOVA without replication can easily be conducted using Excel (Sect. 3.4) and R (Sect. 3.5). (For handling other types of ANOVA with R, we refer the readers to Crawley (Statistics: an introduction using R, 2nd edn. Wiley, Chichester, 2015), Chapter 8.) Finally, it describes how a confidence interval for each system can be constructed based on the data from the first two types of ANOVA (Sect. 3.6).

One-way ANOVA is applicable for comparing (say) m systems using m different user groups; moreover, we shall use one-way ANOVA to discuss *topic set size design* in Chap. 6. Two-way ANOVA without replication is applicable when comparing (says) m systems with the same topic set. However, the reader should be aware that the question addressed with ANOVA is: "are all the populations means equal or not?" It does not tell where the differences lie. If the researcher is interested in the difference between every system pair, then ANOVA is not the test you want: instead, consider a *multiple comparison procedure* such as the *randomised Tukey HSD (honestly significant difference) test* (see Chap. 4).

© Springer Nature Singapore Pte Ltd. 2018

T. Sakai, *Laboratory Experiments in Information Retrieval*,
The Information Retrieval Series 40, https://doi.org/10.1007/978-981-13-1199-4_3

Keywords Confidence intervals · One-way ANOVA · Two-way ANOVA

3.1 One-Way ANOVA

Let us first discuss the simplest form of ANOVA, namely, one-way ANOVA, which considers only one *factor* in an experiment. In IR, the factor is often the choice of IR systems or algorithms: in this book, we consider $m(> 2)$ different systems. With each system i ($i = 1, \ldots, m$), we observe n_i data points such as nDCG scores, x_{ij} ($j = 1, \ldots, n_i$), which we assume are all independent of one another. As was mentioned earlier, this generalises the unpaired t-test situation by considering more than two systems.

Section 3.1.1 discusses one-way ANOVA with *equal* group sizes, where $n_1 = \cdots = n_i = n$, as this is the basis for ANOVA-based *topic set size design* that we discuss in Chap. 6. In that chapter, we shall address the question: "What is the appropriate number of topics n for comparing m systems to ensure (say) 80% statistical power whenever the true difference between the best and the worst systems is (say) 0.1?" Sect. 3.1.2 briefly discusses the more general case with varying n_i, i.e. when the number of observations differs across systems.

3.1.1 One-Way ANOVA with Equal Group Sizes

Table 3.1 shows a situation where one-way ANOVA with equal group sizes is applicable. We have m systems, and for each system, we have n nDCG scores. The data are unpaired: that is, different topic sets were used for evaluating different systems, and the topic set sizes just happened to be the same.

Let us extend the assumption for the two-sample t-test given by Eq. 2.7 (Chap. 2 Sect. 2.3) for handling $m(> 2)$ systems:

$$x_{ij} \sim N(\mu_i, \sigma^2) \,, \tag{3.1}$$

Table 3.1 Data structure for one-way ANOVA with equal group sizes

System	Per-topic scores
1	$x_{11}, x_{12}, \ldots, x_{1n}$
2	$x_{21}, x_{22}, \ldots, x_{2n}$
$\vdots$	$\vdots$
m	$x_{m1}, x_{m2}, \ldots, x_{mn}$

where μ_i is the population mean for System i and σ^2 is the common variance. This homoscedasticity (i.e. equal variances) assumption is essential in ANOVA.[1]

Using an *error term* ε_{ij}, the above model can alternatively be written as:

$$x_{ij} = \mu_i + \varepsilon_{ij}, \quad \varepsilon_{ij} \sim N(0, \sigma^2) . \tag{3.2}$$

Furthermore, let us define the *population grand mean* μ and the i-th *system effect* a_i as follows:

$$\mu = \frac{\sum_{i=1}^{m} \mu_i}{m}, \quad a_i = \mu_i - \mu . \tag{3.3}$$

It can easily be shown that $\sum_{i=1}^{m} a_i = 0$ holds.[2]

The null hypothesis for one-way ANOVA is $H_0 : \mu_1 = \mu_2 = \cdots = \mu_m$ or, equivalently, $a_1 = a_2 = \cdots = a_m = 0$. Thus, we tentatively assume that *all population system means are equal* (to μ); the alternative hypothesis H_1 is that at least one of the system effects a_i is not zero.

From the observed data, we can compute the *sample grand mean* and the *sample system mean* (for $i = 1, \ldots, m$) as follows:

$$\bar{x} = \frac{\sum_{i=1}^{m} \sum_{j=1}^{n} x_{ij}}{mn}, \quad \bar{x}_{i\bullet} = \frac{\sum_{j=1}^{n} x_{ij}}{n} . \tag{3.4}$$

It is easy to see that the following holds for any data:

$$x_{ij} - \bar{x} = (\bar{x}_{i\bullet} - \bar{x}) + (x_{ij} - \bar{x}_{i\bullet}) . \tag{3.5}$$

In words, the difference between each score and the grand mean can be decomposed into (a) the difference between the system mean and the grand mean and (b) the difference between each score and the system mean.

Interestingly, a decomposition analogous to Eq. 3.5 holds true in terms of *sums of squares*: prior to accumulating differences for each of the three difference types expressed in Eq. 3.5, we square each difference so that we will sum up positive values. More specifically, we define the *total sum of squares* as:

$$S_T = \sum_{i=1}^{m} \sum_{j=1}^{n} (x_{ij} - \bar{x})^2 , \tag{3.6}$$

---

[1] See Carterette [3] for a discussion on the validity of the homoscedasticity assumption in the context of IR evaluation.

[2] $\sum_{i=1}^{m} a_i = \sum_{i=1}^{m} (\mu_i - \mu) = \sum_{i=1}^{m} \mu_i - m\mu = 0$ from Eq. 3.3.

which quantifies how the scores vary across the entire data set shown in Table 3.1. Furthermore, let us define the *between-system sum of squares* as:

$$S_A = n \sum_{i=1}^{m} (\bar{x}_{i\bullet} - \bar{x})^2 , \qquad (3.7)$$

which quantifies how the scores vary *across* the rows in Table 3.1. Finally, let the *within-system sum of squares* (for one-way ANOVA) be:

$$S_{E1} = \sum_{i=1}^{m} \sum_{j=1}^{n} (x_{ij} - \bar{x}_{i\bullet})^2 , \qquad (3.8)$$

which quantifies how the scores vary *within* each row in Table 3.1. It can be shown that the following equality holds:

$$S_T = S_A + S_{E1} . \qquad (3.9)$$

That is, the total sum of squares can be decomposed into a between-system sum of squares and a within-system sum of squares.

Since we assume that x_{ij}s are independent from Eq. 3.1 and Theorem 9 (Chap. 1 Sect. 1.2.3), we have:

$$\frac{\sum_{j=1}^{n} (x_{ij} - \bar{x}_{i\bullet})^2}{\sigma^2} \sim \chi^2(n-1) , \qquad (3.10)$$

for $i = 1, \ldots, m$. Since the left side of Eq. 3.10 for a particular i (i.e. system) should also be independent of that for another i, if we sum the left side across all m systems, Theorem 7 (Chap. 1 Sect. 1.2.3) tells us the following:

$$\frac{\sum_{i=1}^{m} \sum_{j=1}^{n} (x_{ij} - \bar{x}_{i\bullet})^2}{\sigma^2} = \frac{S_{E1}}{\sigma^2} \sim \chi^2(\phi_{E1}) , \qquad (3.11)$$

where $\phi_{E1} = m(n-1)$.

We now know what distribution S_{E1} is associated with. What about S_A? Let us first rewrite Eq. 3.1 using Eq. 3.3:

$$x_{ij} \sim N(\mu + a_i, \sigma^2) . \qquad (3.12)$$

Hence, by applying Corollary 1 (Chap. 1 Sect. 1.2.2), we obtain:

$$\bar{x}_{i\bullet} \sim N(\mu + a_i, \sigma^2/n) . \qquad (3.13)$$

Under H_0, the above clearly reduces to:

$$\bar{x}_{i\bullet} \sim N(\mu, \sigma^2/n) .$$ (3.14)

Therefore, under H_0, we can apply Theorem 9 (Chap. 1 Sect. 1.2.3) yet again to obtain:

$$\frac{\sum_{i=1}^{m}(\bar{x}_{i\bullet} - \bar{x})^2}{\sigma^2/n} = \frac{S_A/n}{\sigma^2/n} = \frac{S_A}{\sigma^2} \sim \chi^2(\phi_A) ,$$ (3.15)

where $\phi_A = m - 1$.

From Eqs. 3.11 and 3.15, the definition of F distribution (Chap. 1 Sect. 1.2.5) says that the following holds under H_0:

$$F_0 = \frac{S_A/\sigma^2\phi_A}{S_{E1}/\sigma^2\phi_{E1}} = \frac{S_A/\phi_A}{S_{E1}/\phi_{E1}} \sim F(\phi_A, \phi_{E1}) .$$ (3.16)

That is, if we define the *between-system mean squares* V_A and the *within-system mean squares* V_{E1} as follows,

$$V_A = \frac{S_A}{\phi_A} , \quad V_{E1} = \frac{S_{E1}}{\phi_{E1}} ,$$ (3.17)

then, under H_0,

$$F_0 = \frac{V_A}{V_{E1}} \sim F(\phi_A, \phi_{E1}) .$$ (3.18)

Thus, in essence, F_0 examines how large the between-system mean squares is relative to the within-system mean squares, under H_0. If it is unusually large, then there may be a substantial difference between systems. To be more specific, since F_0 is supposed to obey $F(\phi_A, \phi_{E1})$ under H_0, we reject H_0 if $F_0 \geq F_{inv}(\phi_A, \phi_{E1}; \alpha)$ and say that the system effect is statistically significant at α. The conclusion would be that "there probably is a difference somewhere among the m systems". Figure 3.1 depicts this situation. On the other hand, if $F_0 < F_{inv}(\phi_A, \phi_{E1}; \alpha)$, we say that the system effect is not statistically significant: we cannot conclude whether the system differences are real or not.

3.1.2 One-Way ANOVA with Unequal Group Sizes

Table 3.2 generalises Table 3.1 by allowing each system to have a different number of observations: for example, the i-th system can now have n_i nDCG scores. Since the principle behind one-way ANOVA with unequal group sizes is the same as that discussed in the previous section, let us just discuss how the relevant statistics can be computed when the group sizes differ.

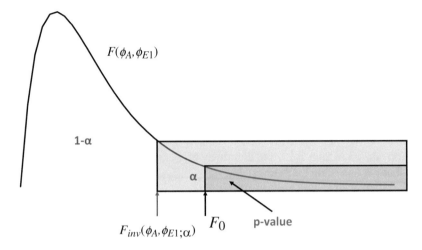

Fig. 3.1 A situation where H_0 is rejected since $F_0 \geq F_{inv}(\phi_A, \phi_{E1}; \alpha)$

Table 3.2 Data structure for
one-way ANOVA with
unequal group sizes

System	Per-topic scores
1	$x_{11}, x_{12}, \ldots, x_{1n_1}$
2	$x_{21}, x_{22}, \ldots, x_{2n_2}$
$\vdots$	$\vdots$
m	$x_{m1}, x_{m2}, \ldots, x_{mn_m}$

Equation 3.6 can now be generalised as:

$$S_T = \sum_{i=1}^{m} \sum_{j=1}^{n_i} (x_{ij} - \bar{x})^2 , \qquad (3.19)$$

and Eq. 3.7 as:

$$S_A = \sum_{i=1}^{m} n_i (\bar{x}_{i\bullet} - \bar{x})^2 , \qquad (3.20)$$

where

$$\bar{x} = \frac{\sum_{i=1}^{m} \sum_{j=1}^{n_i} x_{ij}}{\sum_{i=1}^{m} n_i} , \quad \bar{x}_{i\bullet} = \frac{\sum_{j=1}^{n_i} x_{ij}}{n_i} . \qquad (3.21)$$

Based on the above statistics, the within-system mean squares can now be computed
as (cf. Eq. 3.9):

$$S_{E1} = \sum_{i=1}^{m}\sum_{j=1}^{n_i}(x_{ij} - \bar{x}_{i\bullet})^2 = S_T - S_A . \qquad (3.22)$$

As for the degrees of freedom, let $\phi_A = m - 1$, $\phi_{E1} = \sum_{i=1}^{m} n_i - m$. Note that if $n_i = n$ for $i = 1, \ldots, m$, ϕ_{E1} reduces to $\phi_{E1} = mn - m = m(n - 1)$, which is exactly what we used in one-way ANOVA with equal group sizes. Under the equal means assumption H_0, the following should hold:

$$F_0 = \frac{S_A/\phi_A}{S_{E1}/\phi_{E1}} = \frac{V_A}{V_{E1}} \sim F(\phi_A, \phi_{E1}) . \qquad (3.23)$$

Therefore, reject H_0 iff $F_0 \geq F_{inv}(\phi_A, \phi_{E1}; \alpha)$, as before.

3.2 Two-Way ANOVA Without Replication

Table 3.3 shows a situation where two-way ANOVA without replication is applicable. This situation is very common in laboratory-based IR experiments, where we often have topic-by-run score matrices: that is, the data are paired. Note that x_{i1} ($i = 1, \ldots, m$) represents the scores for the first topic, and so on, and therefore that this setting generalises paired data setting to more than two systems. Here, we regard both systems and topics as factors and hence regard the method discussed here as two-way ANOVA without replication; alternatively, if the observations for each system are regarded as repeated measurements without considering the topic factor, the method is often called *repeated measures ANOVA*. The significance testing procedure is the same in either case.

The model behind two-way ANOVA without replication is as follows:

$$x_{ij} = \mu + a_i + b_j + \varepsilon_{ij} , \quad \varepsilon_{ij} \sim N(0, \sigma^2) . \qquad (3.24)$$

As before, μ is the population grand mean, σ^2 is the common variance, and a_i is the i-th system effect such that $\sum_{i=1}^{m} a_i = 0$. In addition, we now have b_j, the j-th *topic effect* such that $\sum_{j=1}^{n} b_j = 0$. Thus, we assume that the system and topic effects are additive and linearly related to x_{ij} and that the error term ε_{ij} independently obeys the same normal distribution $N(0, \sigma^2)$.

Table 3.3 Data structure for two-way ANOVA without replication

System	Per-topic scores			
1	x_{11}	x_{12}	$\ldots$	x_{1n}
2	x_{21}	x_{22}	$\ldots$	x_{2n}
$\vdots$			$\vdots$	
m	x_{m1}	x_{m2}	$\ldots$	x_{mn}

The null hypothesis for the system effect is $H_0 : a_1 = a_2 = \ldots = a_m = 0$, i.e. there is no system effect at all. The alternative hypothesis H_1 is that at least one of the system effects is not zero.

In addition to the sample grand mean and the sample system mean as defined by Eq. 3.4, we will make use of the *sample topic mean*, defined as:

$$\bar{x}_{\bullet j} = \frac{\sum_{i=1}^{m} x_{ij}}{m} . \tag{3.25}$$

It is easy to see that the following holds for any data:

$$x_{ij} - \bar{x} = (\bar{x}_{i\bullet} - \bar{x}) + (\bar{x}_{\bullet j} - \bar{x}) + (x_{ij} - \bar{x}_{i\bullet} - \bar{x}_{\bullet j} + \bar{x}) . \tag{3.26}$$

Analogously, the following holds for sums of squares:

$$S_T = S_A + S_B + S_{E2} , \tag{3.27}$$

where the total sum of squares S_T and the between-system sum of squares S_A are as defined by Eqs. 3.6 and 3.7, while the *between-topic sum of squares* S_B is defined as

$$S_B = m \sum_{j=1}^{n} (\bar{x}_{\bullet j} - \bar{x})^2 , \tag{3.28}$$

and the *residual sum of squares* S_{E2} is defined as

$$S_{E2} = \sum_{i=1}^{m} \sum_{j=1}^{n} (x_{ij} - \bar{x}_{i\bullet} - \bar{x}_{\bullet j} + \bar{x})^2 . \tag{3.29}$$

From Eqs. 3.9 and 3.27, note that the following holds:

$$S_{E1} = S_B + S_{E2} . \tag{3.30}$$

That is, as two-way ANOVA without replication considers not only the system effect but also the topic effect, the "unexplained part" of the total sum of squares is smaller compared to the case with one-way ANOVA.

Let us define a suite of the mean squares as follows:

$$V_A = \frac{S_A}{\phi_A} , \quad V_B = \frac{S_B}{\phi_B} , \quad V_{E2} = \frac{S_{E2}}{\phi_{E2}} . \tag{3.31}$$

where $\phi_A = m - 1, \phi_B = n - 1, and \phi_{E2} = (m - 1)(n - 1)$. Under H_0 for the system effect, it is known that the following holds:

$$F_0 = \frac{V_A}{V_{E2}} \sim F(\phi_A, \phi_{E2}) \,. \tag{3.32}$$

We therefore reject H_0 iff $F_0 \geq F_{inv}(\phi_A, \phi_{E2}; \alpha)$ and say that the system effect is statistically significant at α, i.e. there is a difference somewhere among the m systems.

If the topic effect is also of interest (which is not the case with repeated measures ANOVA), it can be tested similarly. In this case, the null hypothesis is $H_0 : b_1 = b_2 = \ldots = b_n = 0$, and the test statistic is

$$F_0 = \frac{V_B}{V_{E2}} \sim F(\phi_B, \phi_{E2}) \,. \tag{3.33}$$

We reject H_0 iff $F_0 \geq F_{inv}(\phi_B, \phi_{E2}; \alpha)$.

3.3   Two-Way ANOVA with Replication

Here, two-way ANOVA *with* replication with equal group sizes is briefly discussed.[3] Table 3.4 shows a situation where this test is applicable, with r measurements per cell. Hence the total number of observations is given by mnr.

The model behind two-way ANOVA with replication is as follows:

$$x_{ijk} = \mu + a_i + b_j + (ab)_{ij} + \varepsilon_{ijk} \,, \quad \varepsilon_{ijk} \sim N(0, \sigma^2) \,, \tag{3.34}$$

where $(ab)_{ij}$ represents the *interaction* between System i and Topic j. That is, based on the repeated measurements per cell, we can now consider the possibility that some topics boost the performance of some systems but not others, for example.

Given the repeated measurements as represented by Table 3.4, we first define the following sample means:

Table 3.4 Data structure for two-way ANOVA with replication

System	Per-topic scores			
1	$x_{111}, \ldots, x_{11r}$	$x_{121}, \ldots, x_{12r}$	$\ldots$	$x_{1n1}, \ldots, x_{1nr}$
2	$x_{211}, \ldots, x_{21r}$	$x_{221}, \ldots, x_{22r}$	$\ldots$	$x_{2n1}, \ldots, x_{2nr}$
$\vdots$		$\vdots$		
m	$x_{m11}, \ldots, x_{m1r}$	$x_{m21}, \ldots, x_{m2r}$	$\ldots$	$x_{mn1}, \ldots, x_{mnr}$

---

[3]While two-way ANOVA with replication with unequal group sizes is also possible, this is beyond the scope of this book.

$$\bar{x} = \frac{\sum_{i=1}^{m} \sum_{j=1}^{n} \sum_{k=1}^{r} x_{ijk}}{mnr} \; , \quad \bar{x}_{ij} = \frac{\sum_{k=1}^{r} x_{ijk}}{r} \; , \tag{3.35}$$

$$\bar{x}_{i\bullet} = \frac{\sum_{j=1}^{n} \sum_{k=1}^{r} x_{ijk}}{nr} \; , \quad \bar{x}_{\bullet j} = \frac{\sum_{i=1}^{m} \sum_{k=1}^{r} x_{ijk}}{mr} \; . \tag{3.36}$$

The total sum of squares is now given by:

$$S_T = \sum_{i=1}^{m} \sum_{j=1}^{n} \sum_{k=1}^{r} (x_{ijk} - \bar{x})^2 \; . \tag{3.37}$$

This can now be broken down as follows:

$$S_T = S_A + S_B + S_{A \times B} + S_{E3} \; , \tag{3.38}$$

where

$$S_A = nr \sum_{i=1}^{m} (\bar{x}_{i\bullet} - \bar{x})^2 \; , \quad S_B = mr \sum_{j=1}^{n} (\bar{x}_{\bullet j} - \bar{x})^2 \; , \tag{3.39}$$

$$S_{A \times B} = r \sum_{i=1}^{m} \sum_{j=1}^{n} (\bar{x}_{ij} - \bar{x}_{i\bullet} - \bar{x}_{\bullet j} + \bar{x})^2 \; , \tag{3.40}$$

$$S_{E3} = \sum_{i=1}^{m} \sum_{j=1}^{n} \sum_{k=1}^{r} (x_{ijk} - \bar{x}_{ij})^2 \; . \tag{3.41}$$

Let us define a suite of mean squares as follows:

$$V_A = \frac{S_A}{\phi_A} \; , \quad V_B = \frac{S_B}{\phi_B} \; , \quad V_{A \times B} = \frac{S_{A \times B}}{\phi_{A \times B}} \; , \quad V_{E3} = \frac{S_{E3}}{\phi_{E3}} \; , \tag{3.42}$$

where $\phi_A = m - 1, \phi_B = n - 1, \phi_{A \times B} = (m-1)(n-1),$ and $\phi_{E3} = mn(r-1)$. The test statistic for H_0, "there is no system-topic interaction," is given by:

$$F_0 = \frac{V_{A \times B}}{V_{E3}} \tag{3.43}$$

Reject H_0 (i.e. conclude that there is a statistically significant system-topic interaction) iff $F_0 \geq F_{inv}(\phi_{A \times B}, \phi_{E3}; \alpha)$. Similarly, use the following test statistics for testing the system effect:

$$F_0 = \frac{V_A}{V_{E3}} \; . \tag{3.44}$$

Reject H_0 iff $F_0 \geq F_{inv}(\phi_A, \phi_{E3}; \alpha)$. The topic effect can be tested similarly, using V_B with ϕ_B instead of V_A with ϕ_A.

3.4 Conducting an ANOVA with Excel

Let us conduct one-way ANOVA (with equal group sizes) and two-way ANOVA without replication using http://www.f.waseda.jp/tetsuya/20topics3runs.mat.csv, where we have $m = 3$ systems and $n = 20$ topics. We want to discuss whether there is a statistically significant difference among the three systems.

Figure 3.2 shows how one-way ANOVA can be conducted for the aforementioned data.[4] Note that, since $n = 20, m = 3$ for this example, $\phi_A = 3 - 1 = 2$ and

F25		f_x	=F.DIST.RT(E25,C25,C26)				
	A	B	C	D	E	F	G
1	System1	System2	System3				
2	0.4695	0.3732	0.3575				
3	0.2813	0.3783	0.2435				
4	0.3914	0.3868	0.3167				
5	0.6884	0.5896	0.6024				
6	0.6121	0.4725	0.4766				
7	0.3266	0.233	0.2429				
8	0.5605	0.4328	0.4066				
9	0.5916	0.5073	0.4707				
10	0.4385	0.3889	0.3384				
11	0.5821	0.5551	0.4597				
12	0.2871	0.3274	0.2769				
13	0.5186	0.5066	0.4066				
14	0.5188	0.5198	0.3859				
15	0.5019	0.4981	0.4568				
16	0.4702	0.3878	0.3437				
17	0.329	0.4387	0.2649				
18	0.4758	0.4946	0.4045				
19	0.3028	0.34	0.3253				
20	0.3752	0.4895	0.3205				
21	0.2796	0.2335	0.224				
22							
23		S	phi	V=S/phi	F0	p−value	
24	T	0.726229	59	0.012309			
25	A	0.0753946	2	0.0376973	3.3015235	0.0439858	
26	E1	0.6508345	57	0.0114181			

Fig. 3.2 Conducting one-way ANOVA using Microsoft Excel

---

[4]We pretend here that the data are unpaired as represented in Table 3.1, just as an exercise. In practice, use two-way ANOVA without replication for paired data as represented in Table 3.3.

$\phi_{E1} = 3 * (20 - 1) = 57$. The following summarises a step-by-step approach to conducting ANOVA with Microsoft Excel:

S_T $= \text{DEVSQ(A2:C21)} = 0.726229$ (Eq. 3.6);
S_{E1} $= \text{DEVSQ(A2:A21)} + \text{DEVSQ(B2:B21)} + \text{DEVSQ(C2:C21)} = 0.650834$ (Eq. 3.8);
S_A $= S_T - S_{E1} = 0.075395$ (Eq. 3.9);
V_A $= S_A/\phi_A = 0.075395/2 = 0.037697$ (Eq. 3.17);
V_{E1} $= S_{E1}/\phi_{E1} = 0.650834/57 = 0.011418$ (*ditto*);
F_0 $= V_A/V_{E1} = 3.301524$ (Eq. 3.16);
p-value $= \text{F.DIST.RT}(F_0, 2, 57) = 0.043986$.

Hence the system effect is statistically significant at $\alpha = 0.05$.

For two-way ANOVA without replication, note that $\phi_A = 3 - 1 = 2$ as before, and $\phi_B = 20 - 1 = 19$, $\phi_{E2} = (3-1)*(20-1) = 38$. Also, for example, the mean for System 1 can be obtained as AVERAGE(A2:A21) while the mean for Topic 1 can be obtained as AVERAGE(A2:C2). The following summarises a step-by-step approach to conducting this test for the aforementioned data on Excel:

S_T $= \text{DEVSQ(A2:C21)} = 0.726229$ (Eq. 3.6);
S_A $= 0.075395$ (Eq. 3.7);
S_B $= 0.579826$ (Eq. 3.28);
S_{E2} $= S_T - S_A - S_B = 0.071008$ (Eq. 3.27);
V_A $= S_A/\phi_A = 0.037697$ (Eq. 3.31);
V_{E2} $= S_{E2}/\phi_{E2} = 0.071008/38 = 0.001869$ (Eq. 3.31);
F_0 for the system effect $= V_A/V_{E2} = 20.17365$;
p-value for the system effect $= \text{F.DIST.RT}(F_0, 2, 38) = 1.07012\text{E-}06$.

Similarly, the F_0 for the topic effect is 16.33122; the *p*-value is F.DIST.RT $(F_0, 19, 38) = 8.17317\text{E-}13$. Hence both the system and topic effects are statistically highly significant.

3.5   Conducting an ANOVA with R

Now let us conduct one-way ANOVA and two-way ANOVA without replication for the same data using R. First, load the csv file using the read.table function as shown at the beginning of Chap. 2 Sect. 2.7. Then, try the following R commands:

```
> library(tidyr)
> mattidy <- gather( mat, key=System, value=Score )
> res <- aov( Score ~ factor(System), data=mattidy )
> summary(res)
```

The first command loads the `tidyr` library so that we can utilise the `gather` function, which is useful for rearranging data.[5] The second command converts the 20×3 `mat` data into a two-column data, where the first column represents the systems and the second column represents their scores: `key=System` means "gather all columns from `mat` (i.e. System1, System2, System3) to form a single column called `System`", and `value=Score` means "put the corresponding values in a column called `Score`". The third and the fourth commands above execute one-way ANOVA and display the result.

Figure 3.3 shows what the two-column `mattidy` looks like and the result of one-way ANOVA. Here, `Df` stands for degrees of freedom (ϕ_A and ϕ_{E1}), `Sum Sq` stands for sum of squares (S_A and S_{E1}), and `Mean Sq` stands for mean squares (V_A and V_{E1}); the F-value (F_0) is shown as `F value`, and the p-value is shown as `Pr(>F)`. It can be observed that these results are consistent with the Excel-based ones discussed earlier.

Let us now conduct a two-way ANOVA without replication using the same data. First, try the following R command:

```
> (mat2 <- data.frame(Topic=1:20, mat))
```

This creates a new `data.frame` called `mat2` by adding a Topic ID column to the left of `mat`. Now, try the following commands:

```
> mat2tidy <- gather( mat2, key=System, value=Score,
-Topic)
> res <- aov(Score ~ factor(System) + factor(Topic),
```

```
> library(tidyr)
> mattidy <- gather( mat, key=System, value=Score )
> head(mattidy, 3)
    System  Score
1 System1 0.4695
2 System1 0.2813
3 System1 0.3914
> tail(mattidy, 3)
     System  Score
58 System3 0.3253
59 System3 0.3205
60 System3 0.2240
> res <- aov( Score ~ factor(System), data=mattidy )
> summary(res)
                Df Sum Sq Mean Sq F value Pr(>F)
factor(System)   2 0.0754 0.03770   3.302  0.044 *
Residuals       57 0.6508 0.01142
---
Signif. codes:  0 '***' 0.001 '**' 0.01 '*' 0.05 '.' 0.1 ' ' 1
```

Fig. 3.3 Conducting one-way ANOVA with R

---

[5]Instead of `gather` in the `tidyr` library, `melt` in the `reshape2` may be used to achieve the same effect.

```
+ data=mat2tidy)
> summary(res)
```

The above `gather` command is similar to the one used for one-way ANOVA except that we are now starting from a matrix with four columns (i.e. Topic, System1, System2, System3) and the argument includes `-Topic`. This means "gather all columns *except* the Topic column". As can be seen from Fig. 3.4, this command creates a `data.frame` where the first column represents topic IDs, the second column represents system IDs, and the third column represents the score for each topic-system combination. The results of two-way ANOVA without replication shown in Fig. 3.4 should be easy to interpret: the reader should verify that the results are consistent with the aforementioned Excel-based two-way ANOVA without replication results.

3.6   Confidence Intervals for System Means

In Chap. 2 Sect. 2.8, we discussed how to construct a CI for the *difference* between two population means: the MOE (margin of error) was given by Eqs. 2.21 (paired data), 2.23, or 2.24 (unpaired data). However, given more than two systems, we may want to construct a CI for *each population mean*. Based on the underlying model of one-way ANOVA (with equal group sizes) and that of two-way ANOVA without replication, a common MOE can be computed and used to construct a CI for each system: that is, for System i with sample mean $\bar{x}_{i\bullet}$, the $100(1 - \alpha)\%$ CI is given by $[\bar{x}_{i\bullet} - MOE, \bar{x}_{i\bullet} + MOE]$. From the one-way ANOVA model, the MOE can be computed as [7]:

```
> mat2tidy <- gather( mat2, key=System, value=Score, -Topic )
> head(mat2tidy, 3)
  Topic  System   Score
1      1 System1 0.4695
2      2 System1 0.2813
3      3 System1 0.3914
> tail(mat2tidy, 3)
   Topic  System   Score
58    18 System3 0.3253
59    19 System3 0.3205
60    20 System3 0.2240
> res <- aov(Score ~ factor(System) + factor(Topic),
+ data=mat2tidy)
> summary(res)
               Df Sum Sq Mean Sq F value   Pr(>F)
factor(System)  2 0.0754 0.03770   20.17 1.07e-06 ***
factor(Topic)  19 0.5798 0.03052   16.33 8.17e-13 ***
Residuals      38 0.0710 0.00187
---
Signif. codes:  0 '***' 0.001 '**' 0.01 '*' 0.05 '.' 0.1 ' ' 1
```

Fig. 3.4 Conducting two-way ANOVA without replication with R

$$MOE = t_{inv}(\phi_{E1}; \alpha)\sqrt{V_{E1}/n} \,, \tag{3.45}$$

where $V_{E1} = S_{E1}/\phi_{E1}$ (See Eq. 3.8) and n is the number of topics; from the two-way ANOVA model, it can be computed similarly using ϕ_{E2} and $V_{E2} = S_{E2}/\phi_{E2}$ (see Eq. 3.29) [5, 6].

Take http://www.f.waseda.jp/tetsuya/20topics3runs.mat.csv as an example yet again. The sample mean of the scores for the first system (i.e. Column A in Fig. 3.2) is $\bar{x}_{1\bullet} = 0.45005$. Based on the one-way ANOVA model, the 95% CI for this system can be computed as follows:

$t_{inv}(\phi_{E1}; 0.05)$ = T.INV.2T(0.05, 57) = 2.002465;
MOE = $t_{inv}(\phi_{E1}; 0.05) * \sqrt{0.011418/20} = 0.047846$;
95%CI = $[0.4022, 0.4979]$.

On the other hand, based on the two-way ANOVA (without replication) model, the 95% CI for the same system can be computed as follows:

$t_{inv}(\phi_{E2}; 0.05)$ = T.INV.2T(0.05, 38) = 2.024394;
MOE = $t_{inv}(\phi_{E2}; 0.05) * \sqrt{0.001869/20} = 0.019568$;
95%CI = $[0.4305, 0.4696]$.

Again, because the latter model removes the between-topic sum of squares from the residual sum of squares of one-way ANOVA (see Eq. 3.30), the latter gives a tighter CI.

It should be noted that the CIs of the population means constructed above are not *simultaneous confidence intervals*. The above approach merely ensures that a CI for System i may capture its population mean about 95% of the time, while another CI for System i' may capture its population mean about 95% of the time; it does *not* mean that these CIs capture their respective population means *simultaneously*. One way to construct simultaneous CIs for population means is to utilise the *studentised maximum modulus distribution* [8], which is beyond the scope of this book. For a recent discussion on the CIs for population means in the context of ANOVA, we refer the reader to Baguley [1]. Simultaneous confidence intervals for the *difference* between each system pair will be covered in Chap. 4 Sect. 4.4.4.

References

1. T. Baguley, Calculating and graphing within-subject confidence intervals for ANOVA. Behav. Res. Methods **44**, 158–175 (2012)
2. D. Banks, P. Over, N.-F. Zhang, Blind men and elephants: Six approaches to TREC data. Inf. Retr. **1**, 7–34 (1999)
3. B. Carterette, Multiple testing in statistical analysis of systems-based information retrieval experiments. ACM TOIS **30**(1), 1–34 (2012)
4. M.J. Crawley, *Statistics: An Introduction Using R*, 2nd edn. (Wiley, Chichester, 2015)
5. S. Ishimura, *Analysis of Variance (in Japanese)* (Tokyo Tosho, Bunkyo, 1992)

6. G.R. Loftus, Using confidence intervals in within-subject designs. Psychonomic Bull. Rev. **1**(4), 476–490 (1994)
7. Y. Nagata, *Introduction to Design of Experiments (in Japanese)* (JUSE Press, Shibuya, 2000)
8. Y. Nagata, M. Yoshida, *Introduction to Multiple Comparison Procedures (in Japanese)* (Scientist Press, Shibuya, 1997)

Chapter 4
Multiple Comparison Procedures

Abstract This chapter first discusses the *familywise error rate* problem (Sect. 4.2), which may arise when a researcher applies statistical significance tests multiple times in an experiment. For example, if the researcher has four experimental systems and is interested in comparing every system pair, it is not advisable to conduct a regular *t*-test six times. This chapter then discusses two approaches to lower the familywise error rate, namely, the widely used but arguably obsolete *Bonferroni correction* (Sect. 4.3) and the more recommendable *Tukey HSD (Honestly Significant Difference) test* (Sect. 4.4). While many *multiple comparison procedures* for suppressing the familywise error rate have been proposed, the above two methods are *parametric*, *single-step* methods (Multiple comparison procedures in which the outcome of one hypothesis test determines what to do next are called *stepwise* methods. In contrast, multiple comparison procedures that can process all hypotheses at the same time are called *single-step* methods.) that are suitable for comparing every system pair (Nagata and Yoshida, Introduction to multiple comparison procedures (in Japanese). Scientist Press, 1997). However, the reader should be aware that the Bonferroni correction has low statistical power when handling many hypotheses. Finally, we discuss a distribution-free, computer-based version of the latter test, known as the *randomised Tukey HSD test* (Carterette, ACM TOIS 30(1):1–34, 2012; Sakai, Evaluation with informational and navigational intents. In: Proceedings of WWW 2012, pp 499–508, 2012), for situations where we have a matrix of scores such as a topic-by-run matrix of nDCG values (Sect. 4.5). The paired randomisation test is also discussed as a special case of this test.

Keywords Bonferroni correction · Comparisonwise error rate · Familywise error rate · Multiple comparison procedures · Randomisation test · Randomised Tukey HSD test · Simultaneous confidence intervals · Tukey HSD test

4.1 Introduction

Chapter 3 discussed ANOVA, which tells us whether there is a difference *somewhere* among the m systems; in this chapter, we discuss exactly which of the $m(m-1)/2$

T. Sakai, *Laboratory Experiments in Information Retrieval*,
The Information Retrieval Series 40, https://doi.org/10.1007/978-981-13-1199-4_4

system pairs are statistically significantly different. While some researchers may feel obliged to conduct ANOVA first and *then* proceed to further tests such as Tukey HSD *only if a statistical significance is found in the ANOVA step*, this is not recommended, since Tukey HSD may actually detect significant differences even if ANOVA does not. On the other hand, if a researcher conducts ANOVA and then (say) Tukey HSD *regardless of the outcome of the ANOVA step*, this would cause another familywise error problem, since this practice also involves multiple tests being conducted independently [5]. In short, if the researcher is interested in the difference between every system pair, then it is correct to skip the ANOVA step and directly conduct the (randomised) Tukey HSD test.

Another popular parametric multiple comparison procedure is *Scheffé's method*, which considers a family of null hypotheses in the form $H_0 : \sum_{i=1}^{m} c_i \mu_i = 0$, where c_i's are constants such that $\sum_{i=1}^{m} c_i = 0$ and μ_i's are the population means. $\sum_{i=1}^{m} c_i \mu_i$'s are called *contrasts*. Note, for example, $H_0 : \mu_i = \mu_{i'}$ can be expressed by letting $c_i = 1, c_{i'} = -1$ and all other constants be zero. Scheffé's method includes ANOVA in its steps, and therefore, contrary to the above discussion on the Tukey's HSD test, ANOVA needs to be conducted. However, this book does not discuss Scheffé's method any further, because it is known that it underperforms the Tukey HSD test in terms of statistical power for pairwise comparisons, even for unequal group sizes (i.e. when the sample sizes differ across systems).[1]

In the context of IR evaluation, Tague-Sutcliffe and Blustein [12] used Scheffé's method with two-way ANOVA without replication to analyse TREC-3 results, after pointing out the problem of the *familywise error rate* (see Sect. 4.2). Carterette [1] described a method involving a multivariate generalisation to the *t* distribution for handling multiple hypotheses that may be correlated with one another; his approach is beyond the scope of this book.

4.2 Familywise Error Rate

Suppose that a restaurant has a wine cellar, in which one in every 20 bottles contains sour wine. Thus, if one bottle is randomly drawn from the cellar and served to the customer, there is a 5% chance of making her angry. If two bottles are independently drawn and served to the same table, the probability that both of them are good is 0.95^2, and therefore the probability that at least one of them is sour is $1 - 0.95^2 = 0.0975$. If ten bottles are independently drawn and served for a small drinking party, the probability that all of them are good is 0.95^{10}, and therefore the probability that at least one of them is sour is $1 - 0.95^{10} = 0.4013$. That is, the chance of ruining the party is about 40% [6]!

---

[1]It was proven relatively recently (in 1984, to be exact) that the Tukey HSD test guarantees that the familywise error rate is no larger than α even for unequal group sizes [6].

The above example suggests that, given scores of more than two systems, there would be a problem if we conduct a t-test for different system pairs independently. Recall Table 1.1 from Chap. 1 Sect. 1.1.3: in each t-test, the probability of Type I error (i.e. rejecting a correct hull hypothesis) is α; let us call this the *comparisonwise error rate*. If t-tests are conducted one after another for k different system pairs, the probability that the correct null hypothesis is accepted for all of the k tests is $(1-\alpha)^k$, and therefore the probability that at least one correct null hypothesis among the k is rejected, which is called the *familywise error rate*, amounts to $1-(1-\alpha)^k$, assuming that the test statistics are independent.

By *family*, we mean the set of hypotheses that we are interested in. If there are $m = 3$ systems, we might be interested in the difference between Systems 1 and 2, and the difference between Systems 1 and 3 and nothing else, if System 1 is a proposed system and Systems 2 and 3 are existing ones. Alternatively, we might be interested in the difference between *every system pair*. Hereafter, we only consider the latter situation: given m systems, we want to discuss $m(m-1)/2$ p-values.

4.3 Bonferroni Correction

4.3.1 Principles and Limitations of the Bonferroni Correction

First, a word of warning about the Bonferroni correction: "The thorny issue of multiple comparisons arises because when we do more than one test we are likely to find 'false positives' at an inflated rate (i.e. by rejecting a true null hypothesis more often than indicated by the value of α). *The old fashioned approach was to use Bonferroni's correction*; in looking up a value for Student's t, you divide your α value by the number of comparisons you have done. [...] *Bonferroni's correction is very harsh and will often throw out the baby with the bathwater.* [...] The modern approach is to use contrasts wherever possible, and where it is essential to do multiple comparisons, then to use the wonderfully named Tukey's honestly significant differences" (Crawley [2], pp. 17–18; italics by author). In short, this book does not recommend the use of the Bonferroni correction for IR system evaluation; it is mentioned only because it is simple and still widely used.

The Bonferroni correction utilises the Bonferroni inequality, given by:

$$Pr\left\{\bigcup_{l=1}^{k} E_l\right\} \leq \sum_{l=1}^{k} Pr\{E_l\} \tag{4.1}$$

for any k events $E_1, E_2, \ldots, E_k$. The meaning of the above inequality should be clear from Fig. 4.1: the area of the union of the three circles shown is never greater than the sum of the area of each circle. Equation 4.1 just expresses the above observation in terms of probabilities.

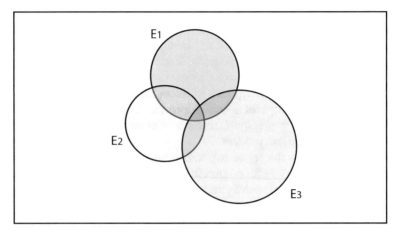

Fig. 4.1 Why the Bonferroni inequality holds for $k = 3$: $Pr\{E_1 \cup E_2 \cup E_3\} \leq Pr\{E_1\} + Pr\{E_2\} + Pr\{E_3\}$

Consider a family of null hypotheses $\mathbf{F}$ where $|\mathbf{F}| = k$. Let $\mathbf{G}(\subseteq \mathbf{F})$ be the set of *correct* null hypotheses, where $|\mathbf{G}| = g(\leq k)$. The Bonferroni correction uses, for every pairwise comparison, α/k instead of α as the significance criterion. That is, it ensures the following for every $H \in \mathbf{G}$:

$$Pr\{H \text{ is rejected}|H \text{ is correct}\} \leq \alpha/k . \tag{4.2}$$

It is clear that the same effect can be achieved by multiplying p-values by k and then comparing them with α.

Using the Bonferroni inequality followed by Eq. 4.2, the familywise error rate probability is given by [5]:

$$Pr\{\text{at least one in } \mathbf{G} \text{ is rejected}|\text{all hypotheses in } \mathbf{G} \text{ are correct}\}$$

$$\leq \quad \sum_{H \in \mathbf{G}} Pr\{H \text{ is rejected}|\text{all hypotheses in } \mathbf{G} \text{ are correct}\}$$

$$= \quad \sum_{H \in \mathbf{G}} Pr\{H \text{ is rejected}|H \text{ is correct}\}$$

$$\leq \quad \sum_{H \in \mathbf{G}} \alpha/k = g\alpha/k \leq k\alpha/k = \alpha . \tag{4.3}$$

4.3.2 Bonferroni Correction with R

Recall `mat` from Chap. 2 Sect. 2.7, our 20×3 topic-by-run matrix. We will also use `mattidy`, the two-column version of `mat` that we created in Chap. 3 Sect. 3.5.

First, let us try conducting a paired t-test for every system pair ($k = m(m - 1)/2 = 3 * 2/2 = 3$ tests done separately). We already looked at the result

```
> t.test(mat$System1, mat$System2, paired = TRUE)

        Paired t-test

data:  mat$System1 and mat$System2
t = 1.3101, df = 19, p-value = 0.2058
alternative hypothesis: true difference in means is not equal to 0
95 percent confidence interval:
 -0.01337109  0.05812109
sample estimates:
mean of the differences
            0.022375

> t.test(mat$System1, mat$System3, paired = TRUE)

        Paired t-test

data:  mat$System1 and mat$System3
t = 8.2607, df = 19, p-value = 1.038e-07
alternative hypothesis: true difference in means is not equal to 0
95 percent confidence interval:
 0.06260099 0.10508901
sample estimates:
mean of the differences
            0.083845

> t.test(mat$System2, mat$System3, paired = TRUE)

        Paired t-test

data:  mat$System2 and mat$System3
t = 4.7726, df = 19, p-value = 0.0001324
alternative hypothesis: true difference in means is not equal to 0
95 percent confidence interval:
 0.03451222 0.08842778
sample estimates:
mean of the differences
             0.06147
```

Fig. 4.2 Conducting multiple paired t-tests independently with R

for Systems 1 and 2, in Chap. 2 Sect. 2.7 Fig. 2.5 (top); Fig. 4.2 shows the results for all three system pairs. Recall that, in a paired t-test, the unbiased variance V_d is computed *for a given system pair* using Eq. 2.4 (Chap. 2 Sect. 2.2), and a t distribution with $\phi = n - 1$ degrees of freedom is used with it. It can be observed that the (unadjusted) p-values are 0.2058, 1.038E-07, and 0.0001324, respectively; since we have $k = 3$ comparisons, the Bonferroni correction multiplies the above values by 3, and the adjusted p-values (to be compared against a standard α) are 0.6174, 3.11E-07, and 0.000397, respectively.

Next, let us treat mat as if it contains three groups of observations that happened to be of the same size and consider three independent *two-sample* t-tests. We assume equal variances and use Student's t-test. We already looked at the result for Systems 1 and 2, in Chap. 2 Sect. 2.7 Fig. 2.5; Fig. 4.3 shows the results for all three system pairs. Recall that, in a Student's t-test, the pooled variance V_p is computed

```
> t.test(mat$System1, mat$System2, var.equal = TRUE)

        Two Sample t-test

data:  mat$System1 and mat$System2
t = 0.6338, df = 38, p-value = 0.53
alternative hypothesis: true difference in means is not equal to 0
95 percent confidence interval:
 -0.04909139  0.09384139
sample estimates:
mean of x mean of y
 0.450050  0.427675


> t.test(mat$System1, mat$System3, var.equal = TRUE)

        Two Sample t-test

data:  mat$System1 and mat$System3
t = 2.3882, df = 38, p-value = 0.02201
alternative hypothesis: true difference in means is not equal to 0
95 percent confidence interval:
 0.01277166 0.15491834
sample estimates:
mean of x mean of y
 0.450050  0.366205


> t.test(mat$System2, mat$System3, var.equal = TRUE)

        Two Sample t-test

data:  mat$System2 and mat$System3
t = 1.998, df = 38, p-value = 0.05291
alternative hypothesis: true difference in means is not equal to 0
95 percent confidence interval:
 -0.0008131775  0.1237531775
sample estimates:
mean of x mean of y
 0.427675  0.366205
```

Fig. 4.3 Conducting multiple Student's t-tests independently with R

for a given system pair with sample sizes n_1 and of n_2 using Eq. 1.29 (Chap. 1 Sect. 1.2.4), and a t distribution with $\phi = n_1 + n_2 - 2$ degrees of freedom is used with it. It can be observed that the (unadjusted) p-values are 0.53, 0.02201, 0.05291, respectively. Therefore, by multiplying them by $k = 3$, the adjusted p-values are computed to be 1, 0.067, 0.159. Note that $0.53 * 3 > 1$, and therefore the adjusted p-value has been set to 1.

The above Bonferroni corrections for the paired t-test and Student's two-sample t-test can also be achieved using the `pairwise.t.test` function available in R. Using the two-column `mattidy` object, try the following:

```
> pairwise.t.test(mattidy$Score, mattidy$System, paired = TRUE,
+ p.adjust.method = "bonferroni")

        Pairwise comparisons using paired t tests

data:  mattidy$Score and mattidy$System

        System1 System2
System2 0.6173   -
System3 3.1e-07 0.0004

P value adjustment method: bonferroni
> pairwise.t.test(mattidy$Score, mattidy$System, pool.sd = FALSE,
+ p.adjust.method = "bonferroni")

        Pairwise comparisons using t tests with non-pooled SD

data:  mattidy$Score and mattidy$System

        System1 System2
System2 1.000    -
System3 0.067    0.159

P value adjustment method: bonferroni
```

Fig. 4.4 Conducting Bonferroni corrections for t-tests with R

```
> pairwise.t.test(mattidy$Score, mattidy$System,
  paired = TRUE,
+ p.adjust.method = "bonferroni")
> pairwise.t.test(mattidy$Score, mattidy$System,
+ pool.sd = FALSE, p.adjust.method = "bonferroni")
```

The results are shown in Fig. 4.4. It can be observed that the function returns the adjusted p-values discussed above.

The reader can also verify that the p-values *without* correction can be obtained using the following commands:

```
> pairwise.t.test(mattidy$Score, mattidy$System,
  paired = TRUE,
+ p.adjust.method = "none")
> pairwise.t.test(mattidy$Score, mattidy$System,
+ pool.sd = FALSE, p.adjust.method = "none")
```

Figure 4.5 shows the results: the first result should be compared with Fig. 4.2 and the second with Fig. 4.3.

When applying multiple Student's t-tests with the Bonferroni correction, it would make sense to utilise a variance pooled by observing the entire topic-by-system matrix instead of computing V_p (Eq. 1.29) by focussing on a particular system pair at a time, since we are assuming equal variances. For example, to compare Systems 1 and 2 with sample sizes n_1 and n_2, instead of using Eq. 2.11 to compute the t value and comparing it against a t-distribution with $\phi = n_1 + n_2 - 2$ degrees of freedom

```
> pairwise.t.test(mattidy$Score, mattidy$System, paired = TRUE,
+ p.adjust.method = "none")

          Pairwise comparisons using paired t tests

data:  mattidy$Score and mattidy$System

        System1 System2
System2 0.20578 -
System3 1e-07    0.00013

P value adjustment method: none
> pairwise.t.test(mattidy$Score, mattidy$System, pool.sd = FALSE,
+ p.adjust.method = "none")

          Pairwise comparisons using t tests with non-pooled SD

data:  mattidy$Score and mattidy$System

        System1 System2
System2 0.530    -
System3 0.022    0.053

P value adjustment method: none
```

Fig. 4.5 Using `pairwise.t.test` without correction

(Eq. 2.11 in Chap. 2 Sect. 2.3), we can compute the following and compare it against a t-distribution with ϕ_{E1} degrees of freedom:

$$T_{12} = \frac{|\bar{x}_{1\bullet} - \bar{x}_{2\bullet}|}{\sqrt{V_{E1}(1/n_1 + 1/n_2)}} , \tag{4.4}$$

where

$$V_{E1} = S_{E1}/\phi_{E1} = \frac{\sum_{i=1}^{m} \sum_{j=1}^{n_i} (x_{ij} - \bar{x}_{i\bullet})^2}{\phi_{E1}} , \quad \phi_{E1} = \sum_{i=1}^{m} n_i - m , \tag{4.5}$$

as defined for one-way ANOVA with unequal group sizes in Chap. 3 Sect. 3.1.2.

Let us re-examine `mattidy`. Since $n_1 = n_2 = n_3 = 20$, $\phi_{E1} = 3 * 20 - 3 = 57$; $V_{E1} = 0.01142$ as shown in Fig. 3.3 (Chap. 3 Sect. 3.5). Using Eq. 4.4 and similar ones for the other system pairs, we obtain $T_{12} = 0.662110, T_{13} = 2.481099, T_{23} = 1.818989$. Hence the corresponding (unadjusted) p-values are $0.510569, 0.016072, 0.074167$.[2] Therefore, the Bonferroni-corrected p-values are $1, 0.048216, 0.222502$.

---

[2]With Microsoft Excel, for example, `T.DIST.2T(0.662110, 57)` $= 0.510569$. With R, try: `2*pt(q=0.66211, df=57, lower.tail=FALSE)`.

```
> pairwise.t.test(mattidy$Score, mattidy$System, pool.sd = TRUE,
+ p.adjust.method = "bonferroni")

        Pairwise comparisons using t tests with pooled SD

data:  mattidy$Score and mattidy$System

        System1 System2
System2 1.000   -
System3 0.048   0.222

P value adjustment method: bonferroni
> pairwise.t.test(mattidy$Score, mattidy$System,
+ p.adjust.method = "bonferroni")

        Pairwise comparisons using t tests with pooled SD

data:  mattidy$Score and mattidy$System

        System1 System2
System2 1.000   -
System3 0.048   0.222

P value adjustment method: bonferroni
```

Fig. 4.6 Conducting Bonferroni corrections for Student's *t*-tests with R (with pooled variances)

The above procedure can also be achieved using `pairwise.t.test`. Try the following two commands:

```
> pairwise.t.test(mattidy$Score, mattidy$System,
  pool.sd = TRUE,
+ p.adjust.method = "bonferroni")
> pairwise.t.test(mattidy$Score, mattidy$System,
+ p.adjust.method = "bonferroni")
```

Figure 4.6 shows the outcome. It can be observed that the results of these two commands are the same. That is, `pairwise.t.test` uses Eq. 4.5 by default. Note that the *p*-values shown in the output are consistent with the aforementioned adjusted *p*-values. It should also be noted that when both `pool.sd` and `paired` are set to TRUE at the same time, the function `pairwise.t.test` returns an error message.

4.4 Tukey HSD Test

Section 4.4.1 describes the Tukey HSD test that is applicable to unequal group sizes, which is often referred to as the *Tukey-Kramer* test. Then, Sect. 4.4.2 discusses the original Tukey HSD test for equal group sizes as a special case of the former. Section 4.4.3 briefly describes the Tukey HSD test with paired observations, which utilises the residual mean squares from two-way ANOVA without replication. Section 4.4.4 introduces *simultaneous confidence intervals* based on the Tukey HSD test. Finally, Sect. 4.4.5 shows how the Tukey HSD test can be conducted using R.

4.4.1   Tukey HSD with Unequal Group Sizes

Table 4.1 replicates Table 3.2 from Chap. 3 Sect. 3.1.2, where we discussed one-way ANOVA with unequal group sizes. Given the above data, we assume that $x_{ij} \sim N(\mu_i, \sigma^2)$ as before. The family of null hypotheses we consider are $\{H_{1,2}, H_{1,3}, \ldots, H_{m-1,m}\}$ where $H_{i,i'}$ means "$\mu_i = \mu_{i'}$"; the alternative hypothesis is $H_{i,i'}^A : \mu_i \neq \mu_{i'}$. We then compute the residual sum of squares and mean squares as follows:

$$ S_{E1} = \sum_{i=1}^{m} \sum_{j=1}^{n_i} (x_{ij} - \bar{x}_{i\bullet})^2 , \quad V_{E1} = \frac{S_{E1}}{\phi_{E1}} , \tag{4.6} $$

where $\bar{x}_{i\bullet} = \sum_{j=1}^{n_i} x_{ij}/n_i$ and $\phi_{E1} = \sum_{i=1}^{m} n_i - m$. Note that this is exactly what we computed for one-way ANOVA with unequal group sizes (Chap. 3 Sect. 3.1.2)[3]; however, recall that this book does *not* recommend conducting an ANOVA prior to a Tukey HSD test. All we need is V_{E1}.

To conduct a Tukey HSD test, we make use of a *studentised range distribution* [6]; let $q_{inv}(m, \phi; P)$ denote the upper $100P\%$ value of this distribution for comparing m systems with ϕ degrees of freedom. In R, this is available as the qtukey function, as we shall discuss in Sect. 4.4.5. For every system pair (i, i'), compute the following test statistic:

$$ t_{i,i'} = \frac{\bar{x}_{i\bullet} - \bar{x}_{i'\bullet}}{\sqrt{V_{E1}(1/n_i + 1/n_{i'})}} , \tag{4.7} $$

and reject $H_{i,i'}$ iff:

Table 4.1 Data structure for one-way ANOVA with unequal group sizes

System	Per-topic scores
1	$x_{11}, x_{12}, \ldots, x_{1n_1}$
2	$x_{21}, x_{22}, \ldots, x_{2n_2}$
$\vdots$	$\vdots$
m	$x_{m1}, x_{m2}, \ldots, x_{mn_m}$

---
[3] Also used in Sect. 4.3.2 in the present chapter in the context of Bonferroni correction.

$$|t_{i,i'}| \geq q_{inv}(m, \phi_{E1}; \alpha)/\sqrt{2} .$$ (4.8)

That is, for every system pair that satisfies Eq. 4.8, the difference is considered statistically significant, while ensuring that the *familywise* error rate is α. Section 4.4.5 discusses how Tukey HSD tests can actually be conducted using R.

The above procedure seems easy to execute provided that we have access to $q_{inv}(m, \phi_{E1}; \alpha)$, the critical value for a studentised range distribution. But what does this test *mean*? The reader should compare Eq. 4.7 with the t-statistic for Student's t-test from Chap. 2 Sect. 2.3, reprinted below:

$$t_0 = \frac{\bar{x}_{1\bullet} - \bar{x}_{2\bullet}}{\sqrt{V_p(1/n_1 + 1/n_2)}}$$ (4.9)

If we conduct $k = m(m-1)/2$ independent Student's t-tests with k different t-statistics, the familywise error rate will suffer, as we have discussed in Sect. 4.2. So, instead, let us consider a null hypothesis distribution for the *maximum* of the k t-statistics, i.e. the statistic that represents the difference between the best and the worst systems given any set of m systems. This null distribution, which is exactly what the studentised range distribution represents, clearly ensures that the comparisonwise Type I error rate for the *largest* between-system difference among the k differences is α. Now, note that we compare all k t-statistics against this studentised range distribution: hence, if the largest among k is not statistically significant, by construction, all other differences would *never* be considered statistically significant either. Thus a familywise error rate of α is ensured.

4.4.2 Tukey HSD with Equal Group Sizes

Table 4.2 replicates Table 3.1 from Chap. 3 Sect. 3.1.1, where we discussed one-way ANOVA with equal group sizes. To conduct a Tukey HSD test for the above data, set up the family of hypotheses as before and compute:

$$S_{E1} = \sum_{i=1}^{m} \sum_{j=1}^{n} (x_{ij} - \bar{x}_{i\bullet})^2 , \quad V_{E1} = \frac{S_{E1}}{\phi_{E1}} ,$$ (4.10)

Table 4.2 Data structure for one-way ANOVA with equal group sizes

System	Per-topic scores
1	$x_{11}, x_{12}, \ldots, x_{1n}$
2	$x_{21}, x_{22}, \ldots, x_{2n}$
$\vdots$	$\vdots$
m	$x_{m1}, x_{m2}, \ldots, x_{mn}$

$\bar{x}_{i\bullet} = \sum_{j=1}^{n} x_{ij}/n$ and $\phi_{E1} = m(n-1)$. The test statistic is:

$$t'_{i,i'} = \frac{\bar{x}_i - \bar{x}_{i'}}{\sqrt{V_{E1}/n}} \, . \tag{4.11}$$

Reject $H_{i,i'}$ iff:

$$|t'_{i,i'}| \geq q_{inv}(m, \phi_{E1}; \alpha) \, . \tag{4.12}$$

Note that the above condition is equivalent to Eq. 4.8 when Eq. 4.7 is rewritten using $n_i = n_{i'} = n$. That is, the procedure described in Sect. 4.4.1 generalises the equal group case described here.

4.4.3 Tukey HSD with Paired Observations

Table 4.3 replicates Table 3.3 from Chap. 3 Sect. 3.2, where we discussed two-way ANOVA without replication given *paired* observations. A version of the Tukey HSD test is available for this setting as well [4]. Set up the family of hypotheses as before and compute:

$$S_{E2} = \sum_{i=1}^{m} \sum_{j=1}^{n} (x_{ij} - \bar{x}_{i\bullet} - \bar{x}_{\bullet j} + \bar{x})^2 \, , \quad V_{E2} = \frac{S_{E2}}{\phi_{E2}} \, , \tag{4.13}$$

where

$$\bar{x} = \frac{\sum_{i=1}^{m} \sum_{j=1}^{n} x_{ij}}{mn} \, , \quad \bar{x}_{i\bullet} = \frac{\sum_{j=1}^{n} x_{ij}}{n} \, , \quad \bar{x}_{\bullet j} = \frac{\sum_{i=1}^{m} x_{ij}}{m} \, , \tag{4.14}$$

and $\phi_{E2} = (m-1)(n-1)$.

The test statistic is:

$$t'_{i,i'} = \frac{\bar{x}_i - \bar{x}_{i'}}{\sqrt{V_{E2}/n}} \, . \tag{4.15}$$

Reject $H_{i,i'}$ iff:

Table 4.3 Data structure for two-way ANOVA without replication

System	Per-topic scores			
1	x_{11}	x_{12}	...	x_{1n}
2	x_{21}	x_{22}	...	x_{2n}
⋮		⋮		⋮
m	x_{m1}	x_{m2}	...	x_{mn}

$$|t'_{i,i'}| \geq q_{inv}(m, \phi_{E2}; \alpha) \,. \tag{4.16}$$

4.4.4 Simultaneous Confidence Intervals

In Chap. 2 Sect. 2.8, we discussed how to construct a CI for the difference between two population means. If we apply that method for multiple system pairs, a CI for System i may capture its population mean difference about 95% of the time, while a CI for System i' may capture its population mean difference about 95% of the time. It does *not* mean that these CIs capture their respective population mean differences *simultaneously*. If we want the CIs to satisfy the latter requirement about 95% of the time, we can construct *simultaneous confidence intervals* for each difference based on the Tukey HSD test.

Based on the Tukey HSD test with unequal group sizes (Sect. 4.4.1), the margin of error can be obtained as:

$$MOE = \frac{q_{inv}(m, \phi_{E1}; \alpha)}{\sqrt{2}} \sqrt{V_{E1}(1/n_i + 1/n_{i'})} \,, \tag{4.17}$$

where V_{E1} is defined as in Eq. 4.6. For the special case with equal group sizes (Sect. 4.4.2), the above MOE reduces to:

$$MOE = q_{inv}(m, \phi_{E1}; \alpha) \sqrt{\frac{V_{E1}}{n}} \,, \tag{4.18}$$

where V_{E1} is defined as in Eq. 4.10. Similarly, for paired observations, the MOE can be obtained using V_{E2} from Sect. 4.4.3 (Eq. 4.13):

$$MOE = q_{inv}(m, \phi_{E2}; \alpha) \sqrt{\frac{V_{E2}}{n}} \,. \tag{4.19}$$

In each case, the $100(1 - \alpha)\%$ CI for the difference between System i and System i' is given by $[\bar{x}_{i\bullet} - \bar{x}_{i'\bullet} - MOE, \ \bar{x}_{i\bullet} - \bar{x}_{i'\bullet} + MOE]$.

4.4.5 Tukey HSD with R

At the time of this writing, Microsoft Excel does not seem to have a function for $q_{inv}(m, \phi; P)$ based on the studentised range distribution. In contrast, it is easy to conduct the Tukey HSD test using R and to obtain a p-value for every system pair while ensuring that the familywise error rate is α.

```
> TukeyHSD(aov(Score ~ factor(System), data = mattidy))
  Tukey multiple comparisons of means
    95% family-wise confidence level

Fit: aov(formula = Score ~ factor(System), data = mattidy)

$`factor(System)`
                     diff        lwr          upr       p adj
System2-System1 -0.022375 -0.1036897  0.058939676 0.7862427
System3-System1 -0.083845 -0.1651597 -0.002530324 0.0418683
System3-System2 -0.061470 -0.1427847  0.019844676 0.1725122
```

Fig. 4.7 Conducting a Tukey HSD test with unpaired observations with R

Let us reuse `mattidy` that we created in Chap. 3 Sect. 3.5, where we considered a 20×3 topic-by-run matrix. A Tukey HSD test with equal group sizes (Sect. 4.4.2) can simply be conducted as follows:

```
> TukeyHSD(aov(Score ~ factor(System), data =
  mattidy))
```

Figure 4.7 shows the result. Since there are $m = 3$ systems, there are three comparisons; the rightmost columns show the p-values. It can be observed that only the difference between Systems 1 and 3 is statistically significant at $\alpha = 0.05$; this fact is also shown in the form of a simultaneous CI $[-0.1651597, -0.002530324]$ (see Eq. 4.17), which does not include zero.

The above simultaneous CI is computed using Eq. 4.17: to verify this, let us use the R function `qtukey`:

$$q_{inv}(m, \phi_{E1}; \alpha) = \text{qtukey}(0.05, \text{nmeans} = 3, \text{df} = 57,$$
$$\text{lower.tail=FALSE}) = 3.403189;$$
$$V_{E1} = 0.011418 \text{ (Sect. 3.4)};$$
$$MOE = 3.403189 * \sqrt{0.011418/20} = 0.081314.$$

Since $\bar{x}_{3\bullet} - \bar{x}_{1\bullet} = -0.0383845$, the simultaneous CI for the difference between Systems 1 and 3 is given by -0.083845 ± 0.081314. Compare this with the lower and upper limits for Systems 1 and 3 in Fig. 4.7.

Next, let us use `mat2tidy` from Chap. 3 Sect. 3.5 to conduct a Tukey HSD test with paired observations, so that we can utilise V_{E2} which is generally smaller than V_{E1}.

```
> TukeyHSD(aov(Score ~ factor(System) + factor(Topic),
+ data=mat2tidy), "factor(System)")
```

The above command examines the system effect only, by having `"factor (System) "` as the second argument of `TukeyHSD`. Figure 4.8 shows the result. It can be observed that the p-values obtained are smaller than those shown in Fig. 4.7 and that this time we have two statistically significantly different pairs.

As before, let us check the simultaneous CI for the difference between Systems 1 and 3, which is $[-0.11718331, -0.05050669]$:

```
> TukeyHSD(aov(Score ~ factor(System) + factor(Topic),
+ data=mat2tidy), "factor(System)")
  Tukey multiple comparisons of means
    95% family-wise confidence level

Fit: aov(formula = Score ~ factor(System) + factor(Topic), data = mat2tidy)

$`factor(System)`
                     diff         lwr         upr      p adj
System2-System1 -0.022375 -0.05571331  0.01096331 0.2428822
System3-System1 -0.083845 -0.11718331 -0.05050669 0.0000011
System3-System2 -0.061470 -0.09480831 -0.02813169 0.0001829
```

Fig. 4.8 Conducting a Tukey HSD test with paired observations with R

$$
\begin{aligned}
q_{inv}(m, \phi_{E2}; \alpha) &= \texttt{qtukey(0.05, nmeans = 3, df = 38,} \\
&\quad \texttt{lower.tail=FALSE)} = 3.4449021; \\
V_{E2} &= 0.001869 \ (\text{Sect. 3.4}); \\
MOE &= 3.4449021 * \sqrt{0.001869/20} = 0.033342.
\end{aligned}
$$

Since $\bar{x}_{3\bullet} - \bar{x}_{1\bullet} = -0.0383845$, the simultaneous CI for the difference between Systems 1 and 3 is given by -0.083845 ± 0.033342. Compare this with the lower and upper limits for Systems 1 and 3 in Fig. 4.8.

Finally, a remark on how the `TukeyHSD` function in R takes the output of `aov` as its input: it seems to imply that ANOVA should be done prior to conducting a TukeyHSD. However, as I stressed at the beginning of this chapter, if the researcher is interested in the difference between every system pair and plans to utilise the Tukey HSD test, then she should use this test directly, without considering ANOVA. In the steps described above, I recommend the reader to view the `aov` function as a tool just for computing the residual mean squares V_{E1} or V_{E2} (with ϕ_{E1} or ϕ_{E1} degrees of freedom), and *not* for an F test that discusses the equality of all population means.

4.5 Randomisation Test and Its Tukey HSD Version

The Tukey HSD test is a parametric, single-step multiple comparison procedure that ensures that the familywise error rate is α. This section describes a *randomised test* version of the Tukey HSD test with paired observations, which is free from the *normality* and *homoscedasticity* assumptions and from the somewhat inconvenient studentised range distribution. In other words, the randomised Tukey HSD test (hereafter referred to as the RTHSD test) is a *distribution-free* version of the Tukey HSD test. First, we discuss the randomisation test for a pair of systems in Sect. 4.5.1. We then generalise the test for handling $m(>2)$ systems, which is exactly the RTHSD test, in Sect. 4.5.2. Both of these tests are also free from the assumption that the observed values are a random sample of a population.

4.5.1 Randomisation Test for Two Systems

For the IR community, Savoy [10] and Sakai [7] have advocated the use of another distribution-free test, called the *bootstrap* test. This test involves *resampling with replacement* from the original topic set to form a null hypothesis distribution. Smucker et al. [11] and Urbano et al. [13] have compared classical significance tests, bootstrap and the randomisation tests, for comparing two systems: see Chap. 2 Sect. 2.1. Sakai [9] examined all ACM SIGIR full papers published between 2006 and 2015 and found that, of the 365 papers that used a significance test for comparing two systems, only 15 (4%) used the randomisation test and 2 (1%) used the bootstrap test; the majority (241 papers; 66%) used the paired t-test. Similar results were obtained for the ACM TOIS journal papers. Thus, these computer-based, distribution-free tests are not so popular in the IR community, despite their simplicity and robustness.

Figure 4.9 explains the concept of the randomisation test for comparing two systems with paired data. The top left corner of this figure shows a score matrix **U** for Systems 1 and 2, evaluated with n topics. The sample means of the two systems are denoted by $\bar{x}_{1\bullet}$ and $\bar{x}_{2\bullet}$, respectively. The null hypothesis H_0 is shown in the bottom left corner: we tentatively assume that there is a *single* hidden system, from which the scores were drawn and then were randomly assigned to one of the systems. Under this assumption, given a pair of scores (x_{1j}, x_{2j}) in the j-th row of **U**, (x_{2j}, x_{1j}) must have been equally likely to occur, since it was by chance that the first score was assigned to System 1 and the second to System 2. Hence, under H_0, variants of **U** shown in Fig. 4.9, where each row has been randomly permuted

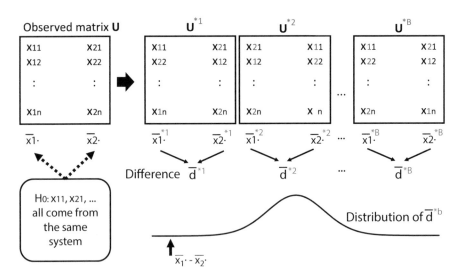

Fig. 4.9 Principle of the randomisation test for paired data

$$\bar{d} = \bar{x}_{1\bullet} - \bar{x}_{2\bullet};$$
$$count = 0;$$
for $b = 1$ **to** B
    for $j = 1$ **to** n
        j-th row of $\mathbf{U}^{*b}$ = a random permutation of the j-th row of $\mathbf{U}$;
    for $i = 1$ **to** 2
        $\bar{x}_{i\bullet}^{*b}$ = mean of the i-th column of $\mathbf{U}^{*b}$;
    $\bar{d}^{*b} = |\bar{x}_{1\bullet}^{*b} - \bar{x}_{2\bullet}^{*b}|$;
    if ($\bar{d}^{*b} \geq |\bar{d}|$) **then** $count$++;
end for
$p\text{-}value = count/B;$

Fig. 4.10 Pseudocode for the paired randomisation test

(i.e. swapped), are as likely to be observed as $\mathbf{U}$. While there are 2^n different ways to create $\mathbf{U}$ and its variants, let us create $B(\ll 2^n)$ variants, $\mathbf{U}^{*1}, \dots, \mathbf{U}^{*B}$ to save computational cost, and compute new sample means $\bar{x}_{1\bullet}^{*b}$ and $\bar{x}_{2\bullet}^{*b}$ from each $\mathbf{U}^{*b}$. The matrices give us B possible values of the between-system difference $\bar{d}^{*b}$, which form our null hypothesis distribution, shown in the bottom right of Fig. 4.9. Compare the actual difference $\bar{d} = \bar{x}_{1\bullet} - \bar{x}_{2\bullet}$ with the null distribution, and count the instances among the B values where $\bar{d}^{*b}$ is more extreme than the observed $\bar{d}$. Figure 4.10 shows a pseudocode for computing the p-value based on the above procedure. Again, note that it does not rely in any way on normality and homoscedasticity assumptions: we are simply permuting the observed data.

A simple implementation of the randomisation test (as a special case of the more general RTHSD test) is available in Sakai's `Discpower` toolkit, available at http://research.nii.ac.jp/ntcir/tools/discpower-en.html.[4] For example, download http://www.f.waseda.jp/tetsuya/20topics2runs.scorematrix.gz, which is actually just the first two columns of http://www.f.waseda.jp/tetsuya/20topics3runs.mat.csv (see Fig. 2.2 in Chap. 2 Sect. 2.6) but with white spaces instead of commas and without the header that contains the system names. We use the `Random-test` script, which should be run on a UNIX-like environment with three arguments, for example[5]:

```
% Random-test 2syslist 20topics2runs.scorematrix 5000
```

Here, the first argument `2syslist` is a file containing a list of system names; the third argument is the number of replicates B. Figure 4.11 shows that the above command creates an output file called `20topics2runs.scorematrix.pvalues.5000` and that it contains the names of the systems, the mean difference, and the

---

[4] Randomisation tests (a.k.a. permutation tests) are also available in the `coin` and `lmPerm` libraries of R.

[5] This may take a while, depending on your computer environment. The reader may choose to implement a more efficient version of this computer-based test.

```
$ cat 20topics2runs.scorematrix
0.4695 0.3732
0.2813 0.3783
0.3914 0.3868
0.6884 0.5896
0.6121 0.4725
0.3266 0.2330
0.5605 0.4328
0.5916 0.5073
0.4385 0.3889
0.5821 0.5551
0.2871 0.3274
0.5186 0.5066
0.5188 0.5198
0.5019 0.4981
0.4702 0.3878
0.3290 0.4387
0.4758 0.4946
0.3028 0.3400
0.3752 0.4895
0.2796 0.2335
$ cat 2syslist
System1
System2
$ Random-test 2syslist 20topics2runs.scorematrix 5000
created 20topics2runs.scorematrix.pvalues.5000
$ cat 20topics2runs.scorematrix.pvalues.5000
System1 System2 0.022375 0.2018
```

Fig. 4.11 Conducting a randomisation test with the Random-test script included in the Discpower toolkit

p-value, which is 0.2018; this is quite close to the p-value we obtained with the paired t-test in Chap. 2 Sect. 2.6, namely, 0.2058.[6]

Table 4.4 shows the effect of increasing the number of replicates B for this particular data set, starting from $B = 5,000$. It can be observed that the p-value does not quite stabilise even at $B = 1,000,000$; however, two decimal places should be enough for discussing the statistical significance of any between-system

---

[6]It was Fisher who discussed the randomisation test to demonstrate the robustness of Student's t-test for non-normal applications [3].

Table 4.4 The effect of increasing B on the accuracy of the randomisation-test p-value (20 topics)

B	p-value
5,000	0.2018
10,000	0.2016
20,000	0.2030
50,000	0.2060
100,000	0.2059
200,000	0.2043
500,000	0.2041
1,000,000	0.2040

difference in most cases. For more discussions on the number of replicates B for the randomisation test, see Smucker, Allan, and Carterette [11].

4.5.2 Randomised Tukey HSD Test

Figure 4.12 explains the concept of the RTHSD test for $m = 3$ systems, where we are interested in the difference between every system pair. We start with an observed matrix **U** shown in the top left corner; note that the data structure is the same as Fig. 4.3 from Sect. 4.4.3. As with the case with $m = 2$ systems, our null hypothesis is that the observed scores actually come from the same systems; thus, if we observe (x_{1j}, x_{2j}, x_{3j}), then other assignments such as (x_{3j}, x_{2j}, x_{1j}) and (x_{2j}, x_{3j}, x_{1j}) must have been equally likely to occur. Hence we create B replicates of the original matrix, $\mathbf{U}^{*b}$ ($b = 1, \ldots, B$). For each replicate $\mathbf{U}^{*b}$, we compute the new sample means as before, and compute the *maximum* between-system difference $\bar{d}^{*b}$, to build a null hypothesis distribution shown in the bottom right of Fig. 4.12. Note that the above null distribution represents the difference between the best possible and the worst possible systems. Finally, we compare each observed between-system difference with the null distribution. Note that this procedure follows the logic of the original Tukey HSD test to ensure that the familywise error rate is α; the only difference is how we obtain the null distribution.

Figure 4.13 shows a pseudocode for computing the p-value for each system based on the above procedure. The reader should compare this with Fig. 4.10: it is clear that the RTHSD test can be seen as a generalisation of the paired randomisation test, since the maximum difference from each RTHSD replicate reduces to the difference between two systems when there are only $m = 2$ systems. Again, this test is completely free from normality and homoscedasticity assumptions.

The aforementioned `Random-test` script was in fact implemented for RTHSD tests with m systems. Download http://www.f.waseda.jp/tetsuya/20topics3runs. scorematrix.gz, which is exactly http://www.f.waseda.jp/tetsuya/20topics3runs.mat.

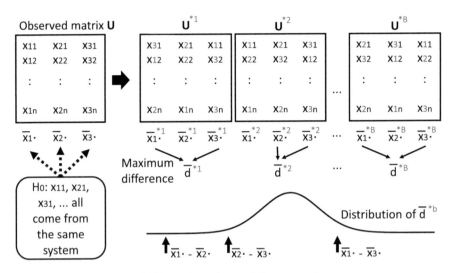

Fig. 4.12 Principle of the RTHSD test for a given topic-by-run matrix

```
foreach system pair (i, i′)
    d̄(i, i′) = x̄ᵢ• − x̄ᵢ′•;
    count(i, i′) = 0;
end for
for b = 1 to B
    for j = 1 to n
        j-th row of U*ᵇ = a random permutation of the j-th row of U;
    for i = 1 to m
        x⃗*ᵇᵢ• = mean of the i-th column of U*ᵇ;
    d*ᵇ = maxᵢ x⃗*ᵇᵢ• − minᵢ x⃗*ᵇᵢ•;
    foreach (i, i′) if ( d*ᵇ ≥ |d̄(i, i′)| ) then count(i, i′)++;
end for
foreach (i, i′)  p-value(i, i′) = count(i, i′)/B;
```

Fig. 4.13 Pseudocode for the randomised Tukey HSD test

csv (see Fig. 2.2 in Chap. 2 Sect. 2.6) but with white spaces instead of commas and
without the header. Then try:

    % Random-test 3syslist 20topics3runs.scorematrix 5000

where the file 3syslist contains a list of system names. As Fig. 4.14 shows, the
above command computes the RTHSD p-value for every system pair: the p-value
for the difference between Systems 1 and 3 is $p = 0.4996$; that between Systems 1
and 3 is $p \approx 0$; and so on.

```
$ cat 20topics2runs.scorematrix
0.4695 0.3732
0.2813 0.3783
0.3914 0.3868
0.6884 0.5896
0.6121 0.4725
0.3266 0.2330
0.5605 0.4328
0.5916 0.5073
0.4385 0.3889
0.5821 0.5551
0.2871 0.3274
0.5186 0.5066
0.5188 0.5198
0.5019 0.4981
0.4702 0.3878
0.3290 0.4387
0.4758 0.4946
0.3028 0.3400
0.3752 0.4895
0.2796 0.2335
$ cat 2syslist
System1
System2
$ Random-test 2syslist 20topics2runs.scorematrix 5000
created 20topics2runs.scorematrix.pvalues.5000
$ cat 20topics2runs.scorematrix.pvalues.5000
System1 System2 0.022375 0.2018
```

Fig. 4.14 Conducting an RTHSD test with the `Random-test` script included in the `Discpower` toolkit

Table 4.5 shows the effect of increasing the number of replicates B for this particular data set, starting from $B = 5,000$. It can be observed that the p-value does not quite stabilise even at $B = 1,000,000$; however, two decimal places should be enough for discussing the statistical significance of any between-system difference in most cases.

Table 4.5 The effect of increasing B on the accuracy of the RTHSD test p-value (20 topics; 3 systems)

	p-value		
B	Systems 1 vs. 2	Systems 1 vs. 3	Systems 2 vs. 3
5,000	0.4996	0	0.0024
10,000	0.4862	0	0.0028
20,000	0.4783	0	0.0020
50,000	0.4763	0.00002	0.0022
100,000	0.4785	0.00001	0.0023
200,000	0.4792	0.000005	0.0023
500,000	0.4792	0.000002	0.0024
1,000,000	0.4791	0.000003	0.0024

References

1. B. Carterette, Multiple testing in statistical analysis of systems-based information retrieval experiments. ACM TOIS **30**(1), 1–34 (2012)
2. M.J. Crawley, *Statistics: an Introduction Using R*, 2nd edn. (Wiley, Chichester, 2015)
3. B. Efron, R.J. Tibshirani, *An Introduction to the Bootstrap* (Chapman & Hall/CRC, Boca Raton, 1993)
4. S. Ishimura, *Analysis of Variance (in Japanese)* (Tokyo Tosho, Bunkyo, 1992)
5. Y. Nagata, How to use multiple comparison procedures (in Japanese). Jpn. J. Appl. Stat. **27**(2), 93–108 (1998)
6. Y. Nagata, M. Yoshida, *Introduction to Multiple Comparison Procedures (in Japanese)* (Scientist Press, Shibuya, 1997)
7. T. Sakai, Evaluating evaluation metrics based on the bootstrap, in *Proceedings of ACM SIGIR*, Seattle, 2006, pp. 525–532
8. T. Sakai, Evaluation with informational and navigational intents, in *Proceedings of WWW, Lyon*, 2012, pp. 499–508
9. T. Sakai, Statistical significance, power, and sample sizes: a systematic review of SIGIR and TOIS, in *Proceedings of ACM SIGIR*, Pisa, 2016, pp. 5–14
10. J. Savoy, Statistical inference in retrieval effectiveness evaluation. Inf. Process. Manag. **33**(4), 495–512 (1997)
11. M.D. Smucker, J. Allan, B. Carterette, A comparison of statistical significance tests for information retrieval evaluation, in *Proceedings of ACM CIKM*, Lisbon, 2007, pp. 623–632
12. J. Tague-Sutcliffe, J. Blustein, A statistical analysis of the TREC-3 data, in *Proceedings of TREC-3*, Gaithersburg, 1995, pp. 385–398
13. J. Urbano, M. Marrero, D. Martín, A comparison of the optimality of statistical significance tests for information retrieval evaluation, in *Proceedings of ACM SIGIR*, 2013, Dublin, pp. 925–928

Chapter 5
The Correct Ways to Use Significance Tests

Abstract Statistical significance testing has been under attack for decades. This section first discusses the criticisms on, and limitations of, significance testing (Sect. 5.1). Then it argues the importance of effect sizes, which typically represent the magnitude of the difference between systems (Sect. 5.2), and finally proposes how researchers should present their significance test results in technical papers and reports (Sect. 5.3). Reporting individual results effectively means that the research community as a whole can accumulate reproducible pieces of evidence and draw general conclusions from them; if researchers adhere to bad practices, that would mean a community where very little is learnt from one another.

Keywords Cohen's d · Dichotomous thinking · Effect sizes · Glass's Δ · Hedge's g

5.1 Limitations of Significance Tests

5.1.1 Criticisms from the Literature

First, let us have a quick look at a small sample of criticisms on the classical significance testing paradigm from literature, arranged chronologically.[1]

> Bakan, 1966 [1]:
> – The test of significance does not provide the information concerning psychological phenomena characteristically attributed to it; and a great deal of mischief has been associated with its use.
> Deming, 1975 [5]:
> – Little advancement in the teaching of statistics is possible, and little hope for statistical methods to be useful in the frightful problems that face man today, until the literature and classroom be rid of terms so deadening to scientific enquiry as null hypothesis,

---

[1] A book edited by Harlow, Mulaik, and Steiger [11] contains a small collection of good arguments *for* and *against* classical significance testing.

© Springer Nature Singapore Pte Ltd. 2018
T. Sakai, *Laboratory Experiments in Information Retrieval*,
The Information Retrieval Series 40, https://doi.org/10.1007/978-981-13-1199-4_5

population (in place of frame), true value, level of significance for comparison of treatments, representative sample.

Loftus, 1991 [22]:
– Despite the stranglehold that hypothesis testing has on experimental psychology, I find it difficult to imagine a less insightful means of transiting from data to conclusions.

Cohen, 1994 [4]:
– And we, as teachers, consultants, authors, and otherwise perpetrators of quantitative methods, are responsible for the ritualization of null hypothesis significance testing (NHST; I resisted the temptation to call it statistical **h**ypothesis **i**nference **t**esting) to the point of meaninglessness and beyond. I argue herein that NHST has not only failed to support the advance of psychology as a science but also has seriously impeded it [bolding by the present author].
– What's wrong with NHST? Well, among many other things, it does not tell us what we want to know, and we so much want to know what we want to know that, out of desperation, we nevertheless believe that it does! What we want to know is Given these data, what is the probability that H_0 is true? But as most of us know, what it tells us is Given that H_0 is true, what is the probability of these (or more extreme) data?

Schmidt, 1996 [30]:
– reliance on statistical significance testing in the analysis and interpretation of research data has systematically retarded the growth of cumulative knowledge in psychology [· · ·].

Rothman, 1998 [26]:
– When writing for **Epidemiology**, you can also enhance your prospects if you omit tests of statistical significance. Despite a wide spread belief that many journals require significance tests for publication, [· · ·] every worthwhile journal will accept papers that omit them entirely. In **Epidemiology**, we do not publish them at all. Not only do we eschew publishing claims of the presence or absence of statistical significance, we discourage the use of this type of thinking in the data analysis [· · ·].

Johnson, 1999 [15]:
– With the hundreds of articles already published that decry the use of statistical hypothesis testing, I was somewhat hesitant about writing another. [· · ·] Our work is important, so we should use the best tools we have available. Rarely, however, is that tool statistical hypothesis testing.

Ziliak and McCloskey, 2008 [34]:
– Statistical significance is surely not the only error in modern science, although it has been, as we will show, an exceptionally damaging one.
– Most important is to minimise Error of the Third Kind, the error of undue inattention, which is caused by trying to solve a scientific problem using statistical significance or insignificance only.

Harlow, 2016 [11]:
– The main opposition to NHST then and now is the tendency for researchers to narrowly focus on making a dichotomous decision to retain or reject a null hypothesis, which is usually not very informative to current or future research in an area [· · ·] Although there is still not a universally agreed upon set of practices regarding statistical inference, there does seem to be more consistency in agreeing on the need to move away from an exclusive focus on NHST [· · ·].

5.1.2 Three Problems, Among Others

The interested reader is referred to the papers mentioned in Sect. 5.1.1 for full discussions on the limitations of statistical significance testing. Here, we only mention three problems with it.

The first problem is that many researchers misinterpret and misuse statistical significance [9]. In March 2016, the American Statistical Association (ASA) published an official statement that really should be obvious to all researchers practicing significance testing, such as [33]:

- P-values do not measure the probability that the studied hypothesis is true, or the probability that the data were produced by random chance alone.
- A p-value, or statistical significance, does not measure the size of an effect or the importance of a result.

Regarding the first point, recall also the second quotation from Cohen [4] (Sect. 5.1.1): researchers often want to know $Pr(H|D)$, i.e. the probability that hypothesis H is true given the observed data D; however, p-value is $Pr(D^+|H)$; the probability of observing D or something more extreme given that H is true.[2] Regarding the second point, *effect sizes* will be discussed in Sect. 5.2.[3]

Another example of common misinterpretation is that for a confidence interval. Does a 95% CI mean "the probability that the population parameter falls into this interval is 95%"? No. In classical statistics, the population parameter is a constant, not a random variable. It is the CIs, constructed from samples, that move about. Thus, as was explained in Chap. 2 Sect. 2.8, a 95% CI means: "If 100 CIs are constructed from 100 different samples, about 95 of them will actually capture the population parameter".

The phrase "statistical significance" is also rather unfortunate, in the sense that some researchers confuse it with *practical* significance. Intentionally or not, the adjective "significant" is often used ambiguously in research papers. For example, in Sakai [28], I reported that, of the 862 SIGIR and TOIS papers that I identified as those deserving but lacking statistical significance testing, as many as 167 (19%) used expressions such as *significant* improvement and *significantly* outperform.

In the long history of IR, several researchers have warned about the difference between statistical significance and *practical* significance. To name but a few:

Sparck Jones, 1974 [31]:
– In a broad way, I shall characterise performance differences, assumed statistically

---

[2] $Pr(H|D)$ can be directly addressed using *Bayesian statistics* [2, 29], but this is beyond the scope of this book; see also Chap. 8.

[3] In his influential paper that advocated the use of parametric tests for IR evaluation, Hull described the p-value as "*a measurement of the probability that the observed difference could have occurred by chance*" [14]. Nooo! On the other hand, he also noted: "*Researchers should simply be cautioned to consider both the statistical significance and the magnitude of the difference*" and thus correctly pointed out the importance of effect size.

significant, as interesting if they are at least noticeable, i.e. of the order of 5–10% different, and as rather more interesting if they are material, i.e. more than 10%.

Sparck Jones, 1981 [32]:
– It must nevertheless be admitted that the basis for applying significance tests to retrieval results is not well established, and it should also be noted that statistically significant performance differences may be too small to be of much operational interest.

Keen, 1992 [16]:
– Table 5.2 has shown that system 7 is better than all others at least at the 5% level of statistical significance on both tests, [· · ·] But is this significance of practical importance? [· · ·] A combination of practical and statistical significance is clearly the best approach.

The second problem of significance testing is often referred to as *dichotomous thinking*: recall the quotations of Harlow from Sect. 5.1.1. The question: *"Is the difference statistically significant or not?"* is not very useful; what is more important is *"How much is the difference? What does that a difference of that magnitude mean to us?"* If a researcher reports in a scientific paper that his proposed method is "statistically significantly better at $\alpha = 0.05$", we cannot know the magnitude or the size of an effect. If a researcher decides not to report on a result just because it is not statistically significant, that may mean missing a practically important discovery. Reporting on the actual p-value is more informative than reporting just the significance criterion α or just putting a few asterisks beside your mean nDCG scores. In Sakai [28], I reported that, of the 565 SIGIR and TOIS papers that I identified as those with significance testing, only 200 (35%) reported either the p-value or the test statistic (from which p-value can be deduced).

The third problem of significance testing is the limited value of the p-value itself, even if it is correctly interpreted (thus solving Problem 1) *and* correctly reported (thus solving Problem 2). More specifically, a p-value is a function of not only the effect size (i.e. magnitude) but also the sample size: a p-value can be made small by using a large sample, and therefore a statistical significance can always be obtained by using a large enough sample. For example, consider the t-statistic from the paired t-test (Chap. 2 Sect. 2.2), replicated here:

$$t_0 = \frac{\bar{d}}{\sqrt{V_d/n}} = \sqrt{n}\frac{\bar{d}}{\sqrt{V_d}} \, . \tag{5.1}$$

It is clear that we can obtain an arbitrarily large t_0 (and hence an arbitrarily small p-value) by making n arbitrarily large. The other component of t_0, namely, $\bar{d}/\sqrt{V_d}$, is a form of *effect size*, which in this case is the magnitude of the between-system difference measured in standard deviation units, known as the *standardised mean difference*. In short, the p-value is not very informative on its own, as it confounds the effect size (i.e. what we are more interested in) with the impact of sample size. Thus, making decisions based solely on p-values is not a good idea.

5.2 Effect Sizes

In Cohen's highly influential book, *effect size* is defined as follows [3]: "it is convenient to use the phrase 'effect size' to mean 'the *degree* to which the phenomenon is present in the population,' or 'the degree to which the null hypothesis is false.' Whatever the manner of representation of a phenomenon in a particular research in the present treatment, the null hypothesis always means that the effect size is zero". More recently, Olejnik and Algina [25] provided the following definition: "An effect-size measure is a standardized index and estimates a parameter that is independent of sample size and quantifies the magnitude of the difference between populations or the relationship between explanatory and response variables." After surveying various definitions of effect size including the above, Kelley and Preacher [17] proposed that the definition be broader and be discussed in the context of a research question: "Effect size is defined as a quantitative reflection of the magnitude of some phenomenon that is used for the purpose of addressing a question of interest."

Below, we describe several specific forms of effect sizes, which may be useful when reported along with statistical significance test results.

5.2.1 Effect Sizes for t-Tests

Let us start with the following form of effect size that may accompany a paired t-test result:

$$d_{paired} = \frac{\bar{d}}{\sqrt{V_d}} = \frac{\sum_{j=1}^{n} d_j / n}{\sqrt{\sum_{j=1}^{n} (d_j - \bar{d})^2 / (n-1)}} , \qquad (5.2)$$

where $d_j = x_{1j} - x_{2j} (j = 1, \ldots, n)$ (Chap. 2 Sect. 2.2). As was discussed in Sect. 5.1.2 of this chapter, d_{paired} is the *standardised mean difference*.[4] From Eq. 5.1, we have:

$$t_0 = \sqrt{n} d_{paired} , \qquad (5.3)$$

in the context of a paired t-test. Hence, if we report d_{paired} together with a p-value of a paired t-test, this is more informative than reporting just the p-value (which is also a function of the sample size n) or reporting just the significance criterion α (e.g. "statistically significant ($p < 0.05$)").

---

[4]As was mentioned earlier, the standardised mean difference measures the effect in standard deviation units. Thus, given the same raw difference $\bar{d}$, the effect size is considered relatively small for high-variance distributions and relatively large for low-variance ones.

Now, let us consider a two-sample situation and discuss a standardised mean difference under the homoscedasticity assumption for Student's t-test given in Eq. 2.7 (Chap. 2 Sect. 2.3). In terms of population parameters, the standardised mean difference, which we are interested in, can be expressed as:

$$\delta = \frac{\mu_1 - \mu_2}{\sigma} . \tag{5.4}$$

This is a *population effect size*. To estimate it from a sample, use the unbiased estimators of μ_1, μ_2, namely, the sample means $\bar{x}_{1\bullet} = \sum_{j=1}^{n_1} x_{1j}/n_1$, $\bar{x}_{2\bullet} = \sum_{j=1}^{n_2} x_{2j}/n_2$ (Chap. 1 Sect. 1.2.4). As for the standard deviation in the denominator, use the following pooled variance, which is an unbiased estimator of the population variance σ^2:

$$V_p = \frac{S_1 + S_2}{n_1 + n_2 - 2} = \left(\sum_{j=1}^{n_1} (x_{1j} - \bar{x}_{1\bullet})^2 + \sum_{j=1}^{n_2} (x_{2j} - \bar{x}_{2\bullet})^2 \right) / (n_1 + n_2 - 2) , \tag{5.5}$$

so that we can obtain a sample effect size given by:

$$g = \frac{\bar{x}_{1\bullet} - \bar{x}_{2\bullet}}{\sqrt{V_p}} . \tag{5.6}$$

This form of effect size is called *Hedge's g* [13]. Since the test statistic for Student's t-test is given by $t_0 = (\bar{x}_{1\bullet} - \bar{x}_{2\bullet})/\sqrt{V_p(1/n_1 + 1/n_2)}$ (Chap. 2 Sect. 2.3), from Eq. 5.6, we obtain:

$$t_0 = \sqrt{\frac{n_1 n_2}{n_1 + n_2}} g . \tag{5.7}$$

Hence, if we report Hedges' g together with a p-value of a Student's t-test, this is more informative than reporting just the p-value or reporting just the significance criterion α (e.g. "statistically significant ($p < 0.05$)").

Probably, *Cohen's d* is more well-known than Hedge's d. However, there appears to be a confusion regarding the exact definition and notation of Cohen's d; sometimes Eq. 5.6 is referred to as Cohen's d. McGrath and Meyer [23] have summarised this confusion in a table, and this book follows their definitions and notations wherever possible. Thus, in contrast to Hedge's g which uses V_p, Cohen's d is defined as:

$$d = \frac{\bar{x}_{1\bullet} - \bar{x}_{2\bullet}}{\sqrt{V'_p}} , \quad V'_p = \frac{S_1 + S_2}{n_1 + n_2} . \tag{5.8}$$

That is, the only difference between Hedge's g and Cohen's d is how the population standard deviation σ is estimated.[5]

Neither $\sqrt{V_p}$ nor $\sqrt{V'_p}$ is an unbiased estimator of σ (even though V_p is an unbiased estimator of σ^2) [24]. However, it is known that $\sqrt{V_p}$ is less biased than $\sqrt{V'_p}$ [24], and therefore Hedge's g (rather than Cohen's d) is generally recommended as an estimator of the population effect size δ (Eq. 5.4) in the context of Student's t-test. Alternatively, a bias-corrected version of Hedge's g may be computed as follows:

$$\hat{\delta} = J(n_1 + n_2 - 2)g \,, \quad J(a) = \frac{\Gamma(a/2)}{\sqrt{a/2}\,\Gamma\left((a-1)/2\right)} \,, \tag{5.9}$$

where $\Gamma(\bullet)$ is the gamma function given by Eq. 1.17 (Chap. 1 Sect. 1.2.3). This estimator is known to be unbiased (i.e. $E(\hat{\delta}) = \delta$) when $n_1 = n_2$.[6] Also, the following approximation of $\hat{\delta}$, which avoids the computationally expensive Gamma function, is available [13, 24]:

$$g_{adj} = \left(1 - \frac{3}{4(n_1 + n_2) - 9}\right)g \,. \tag{5.10}$$

Using a pooled variance such as V_p and V'_p implies that the effect size relies on the homoscedasticity assumption. However, there are cases where we want to avoid this assumption: recall that the paired t-test and Welch's t-test do not rely on homoscedasticity. In such a case, we may want to use the population standard deviation for one of the two samples when discussing Eq. 5.4. For example, if we want to compare a proposed system (which we do not know much about and its standard deviation may not represent a "usual" situation) and a well-known baseline system (i.e. a "control group"), we may choose to use the standard deviation of the baseline system:

$$\Delta = \frac{\bar{x}_{1\bullet} - \bar{x}_{2\bullet}}{\sqrt{V_2}} \,, \quad V_2 = \frac{\sum_{j=1}^{n_2}(x_{2j} - \bar{x}_{2\bullet})^2}{n_2 - 1} \,, \tag{5.11}$$

where V_2 is obtained from the baseline system. The above form of effect size is known as *Glass's* Δ [8]. Also, a biased-corrected version of Glass's Δ is available [10, 24]:

---

[5]To add to the confusion, in a book on effect sizes by Ellis [6], the standard deviation formulas given for Cohen's d and Hedge's g are actually equivalent (pp. 26–27). The difference between Hedge's g and Cohen's d is also discussed in Grissom and Kim [10] (p. 58).
[6]When $n_1 = n_2$, $\hat{\delta}$ is the unique minimum variance unbiased estimator of δ [13].

$$\Delta_{adj} = \left(1 - \frac{3}{4n_2 - 5}\right)\Delta \, . \tag{5.12}$$

In Sakai [29], I used Glass's Δ for comparing the classical significance tests and the corresponding Bayesian tests, as it is applicable to, and comparable across, paired and two-sample t-tests.

Regarding the comparability of effect sizes, it should be noted that d_{paired} (Eq. 5.2) for paired data is *not* an estimator of the population effect size δ as defined by Eq. 5.4. To be more specific, d_{paired} is an estimator of the following population effect size [19, 24]:

$$\delta_{paired} = \frac{\mu_1 - \mu_2}{\sigma\sqrt{2(1 - \rho_{12})}} \, , \tag{5.13}$$

where ρ_{12} is the *population correlation coefficient* between x_{1j}'s ($\sim N(\mu_1, \sigma_1^2)$) and x_{2j}'s ($\sim N(\mu_2, \sigma_2^2)$). Hence d_{paired} should not be compared with (say) Hedge's g obtained from unpaired data.

5.2.2   Effect Sizes for Tukey HSD and RTHSD

In Chap. 4 Sect. 4.4.1, we discussed the comparison of $m(>2)$ systems with $x_{ij} \sim N(\mu_i, \sigma^2)$, with Tukey's HSD test for unequal group sizes, using the residual mean squares:

$$V_{E1} = \frac{S_{E1}}{\phi_{E1}} = \frac{\sum_{i=1}^{m}\sum_{j=1}^{n_i}(x_{ij} - \bar{x}_{i\bullet})^2}{\sum_{i=1}^{m} n_i - m} \, . \tag{5.14}$$

Then in Chap. 4 Sect. 4.4.2, we discussed Tukey's HSD test for equal group sizes as a special case ($n_i = n$ for $i = 1, \ldots, m$, so that $\phi_{E1} = mn - m$). Since Tukey's HSD assumes homoscedasticity, it probably makes sense to consider effect sizes in the form of standardised mean differences (Eq. 5.4) with this test as well. Since V_{E1} is known to be an unbiased estimator of σ^2 (although $\sqrt{V_{E1}}$ is not an unbiased estimator of σ), the following effect size may be useful when reported alongside a p-value for every system pair (i, i'):

$$ES_{E1}(i, i') = \frac{\bar{x}_{i\bullet} - \bar{x}_{i'\bullet}}{\sqrt{V_{E1}}} \, . \tag{5.15}$$

Similarly, in the context of the Tukey HSD with paired observations, we can utilise the residual variance after removing the topic effect:

$$V_{E2} = \frac{S_{E2}}{\phi_{E2}} = \frac{\sum_{i=1}^{m}\sum_{j=1}^{n}(x_{ij} - \bar{x}_{i\bullet} - \bar{x}_{\bullet j} + \bar{x})^2}{(m-1)(n-1)}, \qquad (5.16)$$

where $\bar{x} = (\sum_{i=1}^{m}\sum_{j=1}^{n}x_{ij})/mn$ and $\bar{x}_{\bullet j} = (\sum_{i=1}^{m}x_{ij})/m$ [27]:

$$ES_{E2}(i, i') = \frac{\bar{x}_{i\bullet} - \bar{x}_{i'\bullet}}{\sqrt{V_{E2}}}. \qquad (5.17)$$

These effect sizes may be reported alongside with the p-values of the *randomised* Tukey HSD test as well, although it should be noted that while the RTHSD test does not make any assumptions about the underlying distributions, the above effect sizes assume homoscedasticity. To avoid the homoscedasticity assumption, one may choose to use Glass's Δ instead, if there is a common baseline system among the m systems that can be used to compute an "ordinary" variance V_2 given by Eq. 5.11.

5.2.3 Effect Sizes for ANOVA

In the context of ANOVA, it is common to report effect sizes that represent the contribution of the variance of a particular effect to the overall variance.

Let us first discuss effect sizes based on the one-way ANOVA model (with equal group sizes). Under this model, the following holds for population variances :

$$\sigma_T^2 = \sigma_A^2 + \sigma_{E1}^2, \qquad (5.18)$$

which is analogous to what we observed in Chap. 3 Sect. 3.1 in terms of sum of squares, namely, $S_T = S_A + S_{E1}$. Equation 5.18 says that the total variance can be decomposed into a system variance and a residual variance. Hence, the following measures the contribution of the system variance to the overall variance *if* the population variances are known:

$$\eta^2 = \frac{\sigma_A^2}{\sigma_T^2} = \frac{\sigma_A^2}{\sigma_A^2 + \sigma_{E1}^2}. \qquad (5.19)$$

A simple but biased estimator of η^2 is the following:

$$\hat{\eta}^2 = \frac{S_A}{S_A + S_{E1}}, \qquad (5.20)$$

where, as we have discussed in Chap. 3 Sect. 3.1.1, $S_A = n\sum_{i=1}^{m}(\bar{x}_{i\bullet} - \bar{x})^2$ and $S_{E1} = \sum_{i=1}^{m}\sum_{j=1}^{n}(x_{ij} - \bar{x}_{i\bullet})^2$. However, a more accurate estimator can be obtained by using the following estimates for between-system and within-system population variances [19, 24]:

$$\hat{\sigma}_A^2 = \frac{\phi_A}{mn}(V_A - V_{E1}) , \quad \hat{\sigma}_{E1}^2 = V_{E1} . \tag{5.21}$$

Recall that $\phi_A = m - 1$, $\phi_{E1} = m(n - 1)$. Hence, for one-way ANOVA, by plugging in Eq. 5.21 to Eq. 5.19, we can obtain an accurate estimator of η^2, called $\hat{\omega}^2$ [12, 13, 19]:

$$\begin{aligned}
\hat{\omega}^2 &= \frac{\phi_A(V_A - V_{E1})}{\phi_A(V_A - V_{E1}) + mnV_{E1}} = \frac{S_A - \phi_A V_{E1}}{\phi_A V_A + (mn - m + 1)V_{E1}} \\
&= \frac{S_A - \phi_A V_{E1}}{S_A + m(n - 1)V_{E1} + V_{E1}} = \frac{S_A - \phi_A V_{E1}}{S_A + \phi_{E1} V_{E1} + V_{E1}} \\
&= \frac{S_A - \phi_A V_{E1}}{S_A + S_{E1} + V_{E1}} = \frac{S_A - \phi_A V_{E1}}{S_T + V_{E1}} .
\end{aligned} \tag{5.22}$$

With two-way ANOVA without replication, the total population variance is decomposed as follows:

$$\sigma_T^2 = \sigma_A^2 + \sigma_B^2 + \sigma_{E2}^2 , \tag{5.23}$$

which is analogous to what we observed in Chap. 3, Sect. 3.2 in terms of sum of squares, namely, $S_T = S_A + S_B + S_{E2}$. In this case, we can consider two different population effect sizes for Factor A (i.e. system) that correspond to Eq. 5.19:

$$\eta^2 = \frac{\sigma_A^2}{\sigma_T^2} = \frac{\sigma_A^2}{\sigma_A^2 + \sigma_B^2 + \sigma_{E2}^2} , \tag{5.24}$$

$$\eta_p^2 = \frac{\sigma_A^2}{\sigma_A^2 + \sigma_{E2}^2} . \tag{5.25}$$

The latter is called the *partial* population effect size, which does not include σ_B^2 in the denominator; thus, this eliminates the variance of Factor B (e.g. topic) before measuring the contribution of Factor A (e.g. system) in terms of variances. Simple but biased estimators of the above are given by:

$$\hat{\eta}^2 = \frac{S_A}{S_T} , \quad \hat{\eta}_p^2 = \frac{S_A}{S_A + S_{E2}} , \tag{5.26}$$

where, as we have seen in Chap. 3 Sect. 3.2, $S_{E2} = \sum_{i=1}^{m} \sum_{j=1}^{n} (x_{ij} - \bar{x}_{i\bullet} - \bar{x}_{\bullet j} + \bar{x})^2$. However, more accurate estimators can be obtained by using the following estimates for the population variances [19, 24][7]:

---

[7]The estimates $\hat{\sigma}_A^2$ and $\hat{\sigma}_B^2$ have different forms because while A (system) is a *fixed factor* (i.e. we are interested in a particular set of systems and no other system), B (topic) is considered to

$$\hat{\sigma}_A^2 = \frac{\phi_A}{mn}(V_A - V_{E2}), \quad \hat{\sigma}_B^2 = \frac{1}{m}(V_B - V_{E2}), \quad \hat{\sigma}_{E2}^2 = V_{E2}. \tag{5.27}$$

Recall that $\phi_A = m - 1$, $\phi_B = n - 1$, $\phi_{E2} = (m-1)(n-1)$. Hence, for two-way ANOVA without replication, we have $\hat{\omega}^2$ and $\hat{\omega}_p^2$ as good estimators of $\hat{\eta}^2$ and $\hat{\eta}_p^2$, respectively [19][8]:

$$
\begin{aligned}
\hat{\omega}^2 &= \frac{\phi_A(V_A - V_{E2})}{\phi_A(V_A - V_{E2}) + n(V_B - V_{E2}) + mnV_{E2}} \\
&= \frac{S_A - \phi_A V_{E2}}{S_A + (n-1)V_B + V_B + (mn - (m-1) - n)V_{E2}} \\
&= \frac{S_A - \phi_A V_{E2}}{S_A + S_B + (m-1)(n-1)V_{E2} + V_B} \\
&= \frac{S_A - \phi_A V_{E2}}{S_A + S_B + S_{E2} + V_B} = \frac{S_A - \phi_A V_{E2}}{S_T + V_B},
\end{aligned}
\tag{5.28}
$$

$$\hat{\omega}_p^2 = \frac{\phi_A(V_A - V_{E2})}{\phi_A(V_A - V_{E2}) + mnV_{E2}} = \frac{S_A - \phi_A V_{E2}}{S_A + (mn - \phi_A)V_{E2}}. \tag{5.29}$$

With two-way ANOVA with replication where we can consider interaction, the total population variance is decomposed as follows:

$$\sigma_T^2 = \sigma_A^2 + \sigma_B^2 + \sigma_{A \times B}^2 + \sigma_{E3}^2, \tag{5.30}$$

which is analogous to what we observed in Chap. 3 Sect. 3.4, namely, $S_T = S_A + S_B + S_{A \times B} + S_{E3}$. Hence, the population effect sizes for Factor A are:

$$\eta^2 = \frac{\sigma_A^2}{\sigma_T^2} = \frac{\sigma_A^2}{\sigma_A^2 + \sigma_B^2 + \sigma_{A \times B}^2 + \sigma_{E3}^2}, \tag{5.31}$$

$$\eta_p^2 = \frac{\sigma_A^2}{\sigma_A^2 + \sigma_{E3}^2}, \tag{5.32}$$

and therefore $\hat{\eta}^2$ and $\hat{\eta}_p^2$ are given by:

be a *random factor* (i.e. we could have had a different set of topics); see Kline [19] (Chapter 6, pp. 185–196).

[8]The formula for $\hat{\omega}_p^2$ provided in Okubo and Okada [24] (Chapter 3 Eq. 3.69) contains an error, despite their claim that they substituted Eq. 5.27 into Eq. 5.25 (using a set of notations different from this book). For this reason, Eq. 13 in Sakai [27] also contains an error: as Eq. 5.29 shows, the denominator involves mn, not n.

$$\hat{\eta}^2 = \frac{S_A}{S_T}, \quad \hat{\eta}_p^2 = \frac{S_A}{S_A + S_{E3}}, \tag{5.33}$$

where, as we have seen in Chap. 3 Sect. 3.4, $S_A = nr \sum_{i=1}^{m}(\bar{x}_{i\bullet} - \bar{x})^2$ and $S_{E3} = \sum_{i=1}^{m}\sum_{j=1}^{n}\sum_{k=1}^{r}(x_{ijk} - \bar{x}_{ij})^2$. As before, more accurate estimates can be obtained by using the following estimates for the population variances [19, 24][9]:

$$\hat{\sigma}_A^2 = \frac{\phi_A}{mnr}(V_A - V_{E3}), \quad \hat{\sigma}_B^2 = \frac{\phi_B}{mnr}(V_B - V_{E3}), \tag{5.34}$$

$$\hat{\sigma}_{A \times B}^2 = \frac{\phi_A \phi_B}{mnr}(V_{A \times B} - V_{E3}), \quad \hat{\sigma}_{E3}^2 = V_{E3}. \tag{5.35}$$

Recall that $\phi_A = m-1$, $\phi_B = n-1$, $\phi_{A \times B} = (m-1)(n-1)$, $\phi_{E3} = mn(r-1)$. Hence, for two-way ANOVA with replication, we have [18]:

$$
\begin{aligned}
\hat{\omega}^2 &= \frac{\phi_A(V_A - V_{E3})}{\phi_A(V_A - V_{E3}) + \phi_B(V_B - V_{E3}) + \phi_A\phi_B(V_{A\times B} - V_{E3}) + mnr V_{E3}} \\
&= \frac{S_A - \phi_A V_{E3}}{S_A + S_B + S_{A \times B} + (mnr - (m-1) - (n-1) - (m-1)(n-1))V_{E3}} \\
&= \frac{S_A - \phi_A V_{E3}}{S_A + S_B + S_{A \times B} + mn(r-1)V_{E3} + V_{E3}} \\
&= \frac{S_A - \phi_A V_{E3}}{S_A + S_B + S_{A \times B} + S_{E3} + V_{E3}} = \frac{S_A - \phi_A V_{E3}}{S_T + V_{E3}},
\end{aligned}
\tag{5.36}
$$

$$\hat{\omega}_p^2 = \frac{\phi_A(V_A - V_{E3})}{\phi_A(V_A - V_{E3}) + mnr V_{E3}} = \frac{S_A - \phi_A V_{E3}}{S_A + (mnr - \phi_A)V_{E3}}. \tag{5.37}$$

The effect sizes for Factor B can also be computed using $S_B = mr \sum_{j=1}^{n}(\bar{x}_{\bullet j} - \bar{x})^2$ and ϕ_B instead of S_A and ϕ_A; similarly, those for the interaction $A \times B$ can be computed using $S_{A \times B} = r \sum_{i=1}^{m}\sum_{j=1}^{n}(\bar{x}_{ij} - \bar{x}_{i\bullet} - \bar{x}_{\bullet j} + \bar{x})^2$ and $\phi_{A \times B}$ (see Chap. 3 Sect. 3.4).

Olejnik and Algina [25] propose $\hat{\eta}_G$ and $\hat{\omega}_G$ (where G stands for "generalised") for the purpose of making effect sizes comparable across various research designs.

5.3   How to Report Your Results

The results of a study should be described in sufficient statistical detail that they can be used in planning any further studies of the same hypothesis or of related issues in the same field.

---

[9]Both A and B are treated as fixed factors: see Kline [19] (Chapter 7, p.232).

Results should never be reported only in terms of "NS"[10] or one, two, or three asterisks. These are uninformative symbols, giving no specific information as to what is going on." [20] (p.109).

We already know that reporting p-values is not sufficient.

Effect sizes should be reported with them. Confidence intervals are also more informative than p-values as their widths reflect uncertainty. Visualising CIs also helps us understand the data. Section 5.3.1 provides a few examples of comparing two systems and how the t-test results should be reported; Sect. 5.3.2 discusses the case with more than two systems, where we employ RTHSD described in Chap. 4, Sect. 4.5 and how the results should be reported.[11] Regarding data visualisation, informative representations such as boxplots may also be useful. Details can be found in textbooks on R (e.g. [21]). For generally accepted reporting practices of statistical significance test results, see also Field and Hole [7].

5.3.1 Comparing Two Systems

Suppose that you compared your proposed system (System 1) with a competitive baseline (System 2) using a paired t-test and obtained the result shown at the top of Fig. 5.1: this is a copy of Fig. 2.5 from Chap. 2 Sect. 2.7. First, it would be good to present the sample means of retrieval effectiveness, as shown in Table 5.1. I recommend explicitly stating the sample size n in this way, even if the number is mentioned elsewhere in the paper, because this will facilitate post hoc analyses by other researchers. Furthermore, the information shown in Fig. 5.1 (top) could be reported, with an additional mention of the effect size, as follows. *"We conducted a paired t-test for the difference between our proposed system and the baseline in terms of mean nDCG over n = 20 topics. The difference is not statistically significant (t(19) = 1.3101, p = 0.2058, 95%CI[−0.0134, 0.0581]), with the effect size (standardised mean difference) $d_{paired} = t(19)/\sqrt{n} = 0.2929$. "* Note that the effect size was obtained from the t-value and the sample size by using Eq. 5.3 in this example. Of course, it can also be computed directly using the original definition given by Eq. 5.2 (see Sect. 5.2.1).

Suppose that you compared your proposed system with a competitive baseline using an *unpaired* setting and *Student's* t-test and obtained the results shown at

Table 5.1 Comparison of retrieval effectiveness (sample size: $n = 20$)

System	Mean nDCG
Proposed (System 1)	0.4500
Baseline (System 2)	0.4277

---

[10]Not significant.

[11]While Sakai [27] recommended reporting on an ANOVA results prior to discussing an RTHSD result, we omit the ANOVA step in this book for the reason given at the beginning of Chap. 4.

```
> t.test(mat$System1, mat$System2, paired = TRUE)

        Paired t-test

data:  mat$System1 and mat$System2
t = 1.3101, df = 19, p-value = 0.2058
alternative hypothesis: true difference in means is not equal to 0
95 percent confidence interval:
 -0.01337109  0.05812109
sample estimates:
mean of the differences
              0.022375

> t.test(mat$System1, mat$System2, var.equal = TRUE)

        Two Sample t-test

data:  mat$System1 and mat$System2
t = 0.6338, df = 38, p-value = 0.53
alternative hypothesis: true difference in means is not equal to 0
95 percent confidence interval:
 -0.04909139  0.09384139
sample estimates:
mean of x mean of y
 0.450050  0.427675
```

Fig. 5.1 Conducting a paired t-test and a Student's t-test with R (figure replicated from Chap. 2 Sect. 2.7)

Table 5.2 Comparison of retrieval effectiveness

System	#topics	Mean nDCG
Proposed (System 1)	$n_1 = 20$	0.4500
Baseline (System 2)	$n_2 = 20$	0.4277

the bottom of Fig. 5.1. In this case, the samples for the two systems are different, so the sample sizes should be clearly indicated as shown in Table 5.2, although in our particular example, the two sample sizes are the same. The information shown in Fig. 5.1 (bottom) could be reported, with an additional mention of the effect size, as follows. "*We conducted a Student's t-test for the difference between our proposed system (sample size: $n_1 = 20$) and the baseline (sample size: $n_2 = 20$) in terms of mean nDCG. The difference is not statistically significant ($t(38) = 0.6338$, $p = 0.53$, $95\%CI[-0.0491, 0.0938]$), with the effect size (Hedge's g) $g = t(38)\sqrt{(n_1 + n_2)/(n_1 n_2)} = 0.2004$.* " Note that the effect size was obtained from the t-value and the sample size by using Eq. 5.7 in this example. Of course, it can also be computed directly using the original definition of Hedge's g given by Eq. 5.6 (see Sect. 5.2.1).

If you used Welch's t-test instead of Student's, that means you consciously avoided the homoscedasticity assumption. Therefore, reporting Glass's Δ with the significance test result seems more logical than reporting Hedge's g.

5.3.2 Comparing More Than Two Systems

Suppose you conducted an RTHSD test for $m = 3$ systems with $n = 20$ topics as shown in Fig. 5.2: this is a copy of Fig. 4.14 from Chap. 4 Sect. 4.5.2. Let us report, for every system pair, not only the p-value but also the effect size, denoted ES_{E2}, given by Eq. 4.13 (Sect. 4.4.3) and Eq. 5.17 (Sect. 5.2.2). In Chap. 3 Sect. 3.4, we already obtained $V_{E2} = 0.001869$ from this topic-by-run matrix. Therefore, for example, while the raw difference in sample means between Systems 1 and 2 is 0.0224, the effect size in terms of ES_{E2} is given by $0.0224/\sqrt{0.001869} = 0.5182$. That is, the difference is about half a standard deviation.

Table 5.3 summarises the results of the RTHSD test and the effect sizes; note that information about every system pair is shown. In a research paper, this table could be discussed as follows: "*We conducted a randomised Tukey HSD (RTHSD) test with $B = 5,000$ trials to compare every system pair; the absolute differences and the p-values are summarised in Table 5.3. For example, the mean nDCG of System 1 minus that of System 2 is 0.0224, but this difference is not statistically significant ($p = 0.4996$). It can be observed that System 3 statistically significantly underperforms Systems 1 ($p \approx 0.0000$) and 2 ($p = 0.0024$). Moreover, we computed an effect size (standardised mean difference ES_{E2}) for each system pair, using a standard deviation computed from the following residual sum of squares: $V_{E2} = \sum_{i=1}^{m} \sum_{j=1}^{n} (x_{ij} - \bar{x}_{i\bullet} - \bar{x}_{\bullet j} + \bar{x})^2/((m-1)(n-1)) = 0.001869$, where m is the number of systems (3), n is the number of topics (20), x_{ij} is the nDCG score for System i with Topic j, and $\bar{x}_{i\bullet}, \bar{x}_{\bullet j}, \bar{x}$ are the system mean, topic mean, and the grand mean, respectively. The effect sizes thus computed are also shown in Table 5.3; it can be observed, for example, that Systems 1 and 3 are almost two standard deviations apart from each other.*"[12]

When visualising systems' mean scores, include 95% CIs as error bars, and state clearly that the error bars represent 95% CIs and how the CIs were computed. For example, *Figure 5.3 visualises the mean nDCG scores of Systems 1–3 with 95%*

Table 5.3 Comparison of systems 1–3 in terms of mean nDCG over $n = 20$ topics. In each cell, the difference in mean nDCG, ES_{E2} (effect size) and the p-value (randomised Tukey HSD test with $B = 5,000$ trials), are shown

	System 2		System 3	
System 1	0.0224	$(ES_{E2} = 0.5182)$	0.0838	$(ES_{E2} = 1.9386)$
		$(p = 0.4996)$		$(p \approx 0.0000)$
System 2	–		0.0615	$(ES_{E2} = 1.4227)$
				$(p = 0.0024)$

[12]If the above description of ES_{E2} seems too lengthy, it might be a good idea to just cite this book instead!

```
$ cat 20topics3runs.scorematrix
0.4695 0.3732 0.3575
0.2813 0.3783 0.2435
0.3914 0.3868 0.3167
0.6884 0.5896 0.6024
0.6121 0.4725 0.4766
0.3266 0.2330 0.2429
0.5605 0.4328 0.4066
0.5916 0.5073 0.4707
0.4385 0.3889 0.3384
0.5821 0.5551 0.4597
0.2871 0.3274 0.2769
0.5186 0.5066 0.4066
0.5188 0.5198 0.3859
0.5019 0.4981 0.4568
0.4702 0.3878 0.3437
0.3290 0.4387 0.2649
0.4758 0.4946 0.4045
0.3028 0.3400 0.3253
0.3752 0.4895 0.3205
0.2796 0.2335 0.2240
$ cat 3syslist
System1
System2
System3
$ Random-test 3syslist 20topics3runs.scorematrix 5000
created 20topics3runs.scorematrix.pvalues.5000
$ cat 20topics3runs.scorematrix.pvalues.5000
System1 System2 0.022375 0.4996
System1 System3 0.083845 0
System2 System3 0.06147 0.0024
```

Fig. 5.2 Conducting an RTHSD test with the `Random-test` script included in the `Discpower` toolkit (Figure replicated from Chap. 4 Sect. 4.5.2)

CIs, where the common margin of error is computed as $MOE = t_{inv}((m-1)(n-1); 0.05)\sqrt{V_{E2}/n} = t_{inv}(38; 0.05)\sqrt{0.001869/20} = 0.019568$, where $t_{inv}(\phi; 0.05)$ is the two-sided critical t-value. Note that these CIs are *not* simultaneous CIs (see Chap. 3 Sect. 3.6).

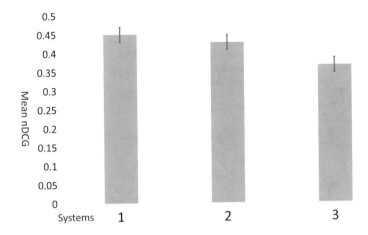

Fig. 5.3 Mean nDCG scores of systems 1–3, with 95% CIs

References

1. D. Bakan, The test of significance in psychological research. Psychol. Bull. **66**(6), 423–437 (1966)
2. B. Carterette, Bayesian inference for information retrieval evaluation, in *Proceedings of ACM ICTIR*, Northampton, 2015, pp. 31–40
3. J. Cohen, *Statistical Power Analysis for the Bahavioral Sciences*, 2nd edn. (Psychology Press, New York, 1988)
4. J. Cohen, The earth is round ($p < .05$). Am. Psychol. **49**(12), 997–1003 (1994)
5. W. Edwards Deming, On probability as a basic for action. Am. Stat. **29**(4), 146–152 (1975)
6. P.D. Ellis, *The Essential Guide to Effect Sizes* (Cambridge University Press, Cambridge/New York, 2010)
7. A. Field, G. Hole, *How to Design and Report Experiments* (Sage Publications, London, 2003)
8. G.V. Glass, B. McGaw, M.L. Smith, *Meta-Analysis in Social Research* (Sage Publications, Beverly Hills, 1981)
9. S. Greenland, S.J. Senn, K.J. Rothman, J.B. Carlin, C. Poole, S.N. Goodman, D.G. Altman, Statistical tests, p values, confidence intervals, and power: a guide to misinterpretations. Eur. J. Epidemiol. **31**(4), 337–350 (2016)
10. R.J. Grissom, J.J. Kim, *Effect Sizes for Research*, 2nd edn. (Routledge, New York, 2012)
11. L.L. Harlow, S.A. Mulaik, J.H. Steiger, *What If There Were No Significance Tests? (Classic Edition)* (Routledge, London, 2016)
12. W.L. Hays, *Statistics (Fifth Edition/International Edition)* (Harcourt Brace College Publishers, Fort Worth, 1994)
13. L.V. Hedges, I. Olkin, *Statistical Methods for Meta-Analysis* (Academic Press, San Diego, 1985)
14. D. Hull, Using statistical testing in the evaluation of retrieval experiments, in *Proceedings of ACM SIGIR'93*, Pittsburgh, 1993, pp. 329–338
15. D.H. Johnson, The insignificance of statistical significance testing. J. Wildlife Manag. **63**(3), 763–772 (1999)
16. E.M. Keen, Presenting results of experimental retrieval comparisons. Inf. Process. Manag. **28**(4), 491–502 (1992)
17. K. Kelley, K.J. Preacher, On effect size. Psychol. Meth. **17**(2), 137–152 (2012)

18. G. Keren, C. Lewis, Partial omega squared for ANOVA designs. Educ. Psychol. Meas. **39**(1), 119–128 (1969)
19. R.B. Kline, *Beyond Significance Testing: Reforming Data Analysis Methods in Behavioral Research* (American Psychology Association, Washington, 2004)
20. H.C. Kraemer, C. Blasey, *How Many Subjects? Statistical Power Analysis in Research*, 2nd edn. (SAGE Publications, Los Angeles, 2016)
21. J.P. Lander, *R for Everyone* (Addison Wesley, Upper Saddle River, 2014)
22. G.R. Loftus, On the tyranny of hypothesis testing in the social sciences. Contemp. Psychol. **36**(2), 102–105 (1991)
23. R.E. McGrath, G.J. Meyer, When effect sizes disagree: the case of r and d. Psychol. Methods **11**(4), 386–401 (2006)
24. M. Okubo, K. Okada, *Psychological Statistics to Tell Your Story: Effect Size, Confidence Interval (in Japanese)* (Keiso Shobo, Bunkyo, 2012)
25. S. Olejnik, J. Algina, Generalized eta and omega squared statistics: measures of effect size for some common research designs. Psychol. Res. **8**(4), 434–447 (2003)
26. K.J. Rothman, Writing for epidemiology. Epidemiology **9**(3), 333–337 (1998)
27. T. Sakai, Statistical reform in information retrieval? SIGIR Forum **48**(1), 3–12 (2014)
28. T. Sakai, Statistical significance, power, and sample sizes: a systematic review of SIGIR and TOIS, in *Proceedings of ACM SIGIR*, Pisa, 2016, pp. 5–14
29. T. Sakai, The probability that your hypothesis is correct, credible intervals, and effect sizes for IR evaluation, in *Proceedings of ACM SIGIR*, Shinjuku, 2017, pp. 25–34
30. F.L. Schmidt, Statistical significance testing and cumulative knowledge in psychology: implications for training of researchers. Psychol. Meth. **1**(2), 115–129 (1996)
31. K. Sparck Jones, Automatic indexing 1974: a state of the art review. Technical report, Computer Laboratory, University of Cambridge, British Library Research and Development Report No. 5193 (1974)
32. K. Sparck Jones, Retrieval system tests 1958–1978, in *Information Retrieval Experiment*, chap. 12, ed. by K. Sparck Jones. (Butterworths, London, 1981)
33. R.L. Wasserstein, N.A. Lazar, The ASA's statement on p-values: context, process, and purpose. Am. Stat. **70**(2), 129–133 (2016)
34. S.T. Ziliak, D.N. McCloskey, *The Cult of Statistical Significance: how the Standard Error Costs us Jobs, Justice, and Lives* (The University of Michigan Press, Ann Arbor, 2008)

Chapter 6
Topic Set Size Design Using Excel

Abstract This chapter discusses *topic set size design*, which enables test collection builders to determine the number of topics to create based on statistical requirements. First, an overview of five topic set size design methods is provided (Sect. 6.1), followed by details on each method (Sects. 6.2, 6.3, 6.4, 6.5, and 6.6). These methods are based on a desired statistical power (for the paired t-test, the two-sample t-test, and one-way ANOVA) or on a desired cap on the expected width of the confidence interval of the difference in means for paired and unpaired data. The simple Excel tools that I devised are based on the *sample size design* techniques as described in Nagata Y (How to design the sample size (in Japanese). Asakura Shoten, 2003). As these methods require an estimate of the population within-system variance for a given evaluation measure (or the variance of the score differences in the case of paired data), this chapter then describes how the variance can be estimated from pilot data (Sect. 6.7). Finally, it discusses the relationship across the different topic set size design methods (Sect. 6.8).

Keywords Confidence intervals · One-way ANOVA · Paired t-test · Sample sizes · Topic set size design · Variance estimates

6.1 Overview of Topic Set Size Design

In Chap. 5 Sect. 5.1.2, we discussed a problem with the p-value, namely, that it is a function of not only the effect size (e.g., standardised mean difference) but also the sample size. There are four factors in statistical significance testing: the Type I error probability α, the Type II error probability β (or the statistical power $(1 - \beta)$), the effect size, and the sample size. Each of these is a function of the other three.

© Springer Nature Singapore Pte Ltd. 2018
T. Sakai, *Laboratory Experiments in Information Retrieval*,
The Information Retrieval Series 40, https://doi.org/10.1007/978-981-13-1199-4_6

Suppose we know the desired α and β and also have an idea of the effect size: how large should our sample be? This is the question we address in this chapter.[1]

A quick history on topic set sizes for IR evaluation: in 1975, Sparck Jones and Van Rijsbergen discussed the requirements for an ideal IR test collection, in which they argued that "*<75 requests are of no real value*", "*250 requests are minimally acceptable*", and "*>1000 requests are needed for some purposes*" [27]. However, these recommendations were probably not based on any specific statistical considerations. In contrast, the subsequent analyses by Sparck Jones and Bates from 1977 [28] and Gilbert and Sparck Jones from 1979 [11] were statistically motivated, under the premise that the sign test was to be used for pairwise comparisons and that the evaluation measures were recall and precision which can be regarded as probabilities. The statistical power of the sign test and a form of effect size, namely, the relative difference between two recall or precision values, were considered there. However, the studies primarily examined the number of documents that need to be judged *given* a topic set size.[2] For example, Gilbert and Sparck Jones remark on topic set sizes as follows: "*Since there is some doubt about the feasibility of getting 1000 requests, or the convenience of such a large set for future experiments, we consider 500 requests.*"

It is interesting that the default topic set size at TREC is strikingly smaller than those considered in the above studies, namely, 50. It appears that the practice of creating 50 topics originates from TREC-1 in 1992, where topics 1–50 were used for the *routing* track and 51–100 were used for the ad hoc track [12]. In 2009, Voorhees [31] reported on a *swap rate* experiment where 100 topics from the TREC 2004 Robust track (originally from TREC-7 and TREC-8) were randomly and repeatedly split in half and conflicts across the two topic sets according to the paired t-test were counted. Her conclusion was: "*Fifty-topic sets are clearly too small to have confidence in a conclusion when using a measure as unstable as P(10).*[3] *Even for stable measures, researchers should remain skeptical of conclusions demonstrated on only a single test collection.*" Related studies prior to her work include Zobel [34], Voorhees and Buckley [32], Sanderson and Zobel [26], and Sakai [20]. In Sect. 6.8, I shall present recommended topic set sizes based on topic set size design with real TREC data: the results suggest that *fifty-topic sets are clearly too small if we want high statistical power for small effects,* although the required topic set sizes vary depending on the variance of the measure chosen. In other words, an IR experiment with 50 topics may often be *underpowered* (see also Chap. 7): we may be overlooking many between-system differences that are *real*.

---

[1]This chapter relies heavily on Nagata's formula derivations for sample size design [16], but the book is in Japanese. For discussions in English on sample sizes power analysis, the reader is referred to Ryan [18], Murphy, Myors, and Wolach [15], and Kraemer and Blasey [14].

[2]Gilbert and Sparck Jones [11] (page A4) do report on a table that shows the required number of topics as a function of the number of relevant or retrieved documents per topic. For example, if the number of relevant documents per topic is five and we want 5% Type I error probability and 95% statistical power with the sign test, 830 topics are required according to their analysis.

[3]Precision at document cuttoff 10.

Regarding topic set sizes of TREC, an obvious exception was the TREC Million Query Tracks that were run from 2007 to 2009 [1, 2, 6, 7]. This track series had about 1,800, 800, and 700 judged topics, respectively. However, these topics were "sparsely judged" [2], by leveraging techniques called the *minimal test collection* [5] and *statAP*[4] methods that can be used to estimate true evaluation measure scores from very few judged documents.

The above Million Query Track techniques for pooling and judging are beyond the scope of this book: we adhere to the standard practice of pooling with a particular *pool depth* (i.e. how many top-ranked documents to take from each submitted run). Thus, a depth-k pool for a particular topic is a union of top k documents for that topic from each run, and the cost for assessing the entire pool increases monotonically with k. Hence, the total assessment cost is the sum of the assessment cost for each pool across n topics; increasing n means increasing the total assessment cost.

Webber, Moffat, and Zobel [33] proposed an incremental method for constructing a topic set for a test collection where a topic is added to the set one by one (and the documents for that topic are judged) until the desired statistical power for the paired t-test is achieved. In contrast, the topic set size design approaches described in this chapter are more straightforward in that they provide simple answers to questions such as: "How many topics should I prepare if I want the paired t-test for any system pair with an effect size of 0.2 to achieve 80% statistical power while ensuring that the Type I error probability is no more than 5%?"

In this chapter, the following five Excel tools for topic set size design are described[5]:

- `samplesizeTTEST2.xlsx` (Sect. 6.2):
  available at http://www.f.waseda.jp/tetsuya/samplesizeTTEST2.xlsx
- `samplesize2SAMPLET.xlsx` (Sect. 6.3):
  available at http://www.f.waseda.jp/tetsuya/samplesize2SAMPLET.xlsx
- `samplesizeANOVA2.xlsx` (Sect. 6.4):
  available at http://www.f.waseda.jp/tetsuya/samplesizeANOVA2.xlsx
- `samplesizeCI2.xlsx` (Sect. 6.5):
  available at http://www.f.waseda.jp/tetsuya/samplesizeCI2.xlsx
- `samplesize2SAMPLECI.xlsx` (Sect. 6.6):
  available at http://www.f.waseda.jp/tetsuya/samplesize2SAMPLECI.xlsx

Hereafter, they will be referred to without the `xlsx` suffix for brevity.

`samplesizeTTEST2` and `samplesize2SAMPLET` are tools for computing the required number of topics in order to achieve a specified statistical power in a paired t-test and a two-sample t-test, respectively; `samplesizeCI2` and `samplesize2SAMPLECI` are for computing the required number of topics given

---

[4]http://www.ccs.neu.edu/home/jaa/papers/drafts/statAP.pdf

[5]These tools are slightly easier to use than their earlier versions, `samplesizeTTEST.xlsx`, `samplesizeANOVA.xlsx`, and `samplesizeCI.xlsx`, in that there is no need for the user to scroll down the Excel sheet to find the right topic set size anymore.

a cap on the expected CI width for the difference between any pair of systems, for paired and unpaired data, respectively. Thus, these tools only consider comparisons of a $m = 2$ systems. While we discussed multiple comparison procedures and simultaneous confidence intervals for comparing $m(>2)$ systems in Chap. 4, there are no simple and well-established sample size design techniques based on these approaches, according to Nagata [17]. One difficulty with discussing statistical power for multiple comparison procedures is how exactly to define the power, since *multiple* null hypotheses are involved. On the other hand, it is relatively easy to consider required topic sizes from the viewpoint of statistical power for one-way ANOVA with m systems, as discussed below.

samplesizeANOVA2 is a tool for computing the required number of topics in order to achieve a specified statistical power in one-way ANOVA for comparing $m(\geq 2)$ systems. While two-way ANOVA without replication is more appropriate for situations where we have an $n \times m$ topic-by-run matrix, this topic is not covered in Nagata [16]. However, sample sizes required by one-way ANOVA can be considered as pessimistic estimates for those required by two-way ANOVA without replication; Erring on the side of oversampling is better than erring on the side of undersampling if the objective is to ensure a certain level of statistical power.

As samplesize2SAMPLET, samplesizeANOVA2, and samplesize2 SAMPLECI are for unpaired data, they require an estimate of the common population variance σ^2, for a given evaluation measure, as input. On the other hand, samplesizeTTEST2 and samplesizeCI2 are for paired data, and they require as input an estimate of the population variance for the score *differences*, given by $\sigma_d^2 = \sigma_1^2 + \sigma_2^2$ (See Chap. 2 Sect. 2.2). How to obtain such estimates from past data is discussed in Sect. 6.7.

In Sect. 6.8, it is shown that one-way ANOVA with $m = 2$ systems is strictly equivalent to the two-sample t-test, and power-based topic set size design results based on the t-tests and one-way ANOVA are compared for $m = 2$. Then, CI-based topic set size design results based on paired and unpaired data are compared. Finally, I shall discuss why one-way ANOVA-based and CI-based topic set size results turn out to be virtually identical under some settings.

6.2   Topic Set Size Design with the Paired t-Test

samplesizeTTEST2 is a tool for computing the required number of topics in order to achieve a specified statistical power in a paired t-test.

6.2.1   How to Use the Paired t-Test-Based Tool

samplesizeTTEST2 requires the following as input:

α Type I Error probability.

β Type II Error probability. That is, you want $100(1 - \beta)\%$ statistical power for any paired *t*-test (see below).

minD_t   *Minimum detectable difference*. That is, whenever the true difference between two systems (in terms of a particular evaluation measure) is *minD_t* or larger, you want to guarantee $100(1 - \beta)\%$ statistical power.

$\hat{\sigma}_d^2$ Variance estimate for the score *differences* in terms of the above evaluation measure. How to obtain such an estimate will be discussed in Sect. 6.7.

By entering the above values into the "From absolute diff" sheet of the Excel file, the topic set size required can easily be obtained. The other sheet, "From effect size", will be discussed in Sect. 6.2.2.

Suppose you are interested in the statistical power for comparing any system pair with a paired *t*-test at $\alpha = 0.05$ with an evaluation measure whose estimated population variance for the score differences is $\hat{\sigma}_d^2 = 0.2$. If you want to guarantee 80% statistical power whenever the true difference between any two systems is 0.1 or larger, enter $(\alpha, \beta, minD_t, \hat{\sigma}_d^2) = (0.05, 0.20, 0.10, 0.20)$ into the "From absolute diff" sheet. Figure 6.1 shows what happens: it can be observed that the required topic set size is 159: this is the minimum topic set size that satisfies the required statistical power under the specified condition.[6]

6.2.2 How the Paired t-Test-Based Topic Set Size Design Works

This section explains how samplesizeTTEST2 obtains the required sample size for the paired *t*-test, given $(\alpha, \beta, minD_t, \hat{\sigma}_d^2)$.

	A	B	C	D	E	F	G
1	alpha	beta	minDelta_t	minD_t	sigma_d^2		
2	0.05	0.2	0.2236	0.1	0.2		
3	zalpha/2	z1−beta	n approx	n	large enough?	w	u <= ? (−w)
4	1.959964	−0.841621	158.9	162	TRUE	1.9748081	−4.78887881
5	README		n recommended	161	TRUE	1.9749016	−4.78002614
6			159	160	TRUE	1.9749962	−4.77114558
7			large enough?	159	TRUE	1.9750921	−4.76223685
8			TRUE	158	FALSE	1.9751892	−4.75329969
9				157	FALSE	1.9752875	−4.74433383
10				156	FALSE	1.9753871	−4.73533897
11				155	FALSE	1.9754881	−4.72631484
12				154	FALSE	1.9755903	−4.71726116
13				153	FALSE	1.9756939	−4.70817763

Fig. 6.1 How to use samplesizeTTEST2: an example (sheet, from absolute diff)

---

[6]The achieved power is computed in Column K, although not shown in Fig. 6.1.

First, let us recall what we do when we conduct the two-sided paired t-test (Chap. 2, Sect. 2.2). We start with the assumptions:

$$x_{1j} \sim N(\mu_1, \sigma_1^2), \quad x_{2j} \sim N(\mu_2, \sigma_2^2) \,, \tag{6.1}$$

and hence:

$$d_j \sim N(\mu_1 - \mu_2, \sigma_d^2) \,, \tag{6.2}$$

where

$$\sigma_d^2 = \sigma_1^2 + \sigma_2^2 \,. \tag{6.3}$$

We have n topics and therefore $\phi = n - 1$. From Corollary 5 (Chap. 1 Sect. 1.2.4), we obtain:

$$t = \frac{\bar{d} - (\mu_1 - \mu_2)}{\sqrt{V_d/n}} \sim t(\phi) \,. \tag{6.4}$$

We set up the null hypothesis $H_0 : \mu_1 = \mu_2$ and the alternative hypothesis $H_1 : \mu_1 \neq \mu_2$; we reject H_0 iff $|t_0| \geq t_{inv}(\phi; \alpha)$. That is, the probability of rejecting H_0 is given by:

$$Pr\{t_0 \leq -t_{inv}(\phi; \alpha)\} + Pr\{t_0 \geq t_{inv}(\phi; \alpha)\} \tag{6.5}$$

$$= Pr\{t_0 \leq -t_{inv}(\phi; \alpha)\} + 1 - Pr\{t_0 \leq t_{inv}(\phi; \alpha)\} \,. \tag{6.6}$$

Now, see Table 6.1: this is a copy of Table 1.1 from Chap. 1 Sect. 1.1.3. The reality is either "H_0 is true" or "H_0 is false" (i.e. H_1 is true). If H_0 is indeed true, then Eq. 6.4 reduces to:

$$t_0 = \frac{\bar{d}}{\sqrt{V_d/n}} \sim t(\phi) \,, \tag{6.7}$$

and Eq. 6.6 is exactly α; in fact, this property constitutes the very definition of $t_{inv}(\phi; \alpha)$.

Table 6.1 α, β and statistical power

	We cannot reject H_0; (not statistically significant)	We reject H_0 (statistically significant)
H_0 is true (e.g. systems are equivalent)	correct conclusion (probability, $1 - \alpha$)	Type I error (probability, α)
H_0 is false (e.g. systems are not equivalent)	Type II error (probability, β)	Correct conclusion (probability, $1 - \beta$)

On the other hand, *if H_0 is false*, since $\mu_1 \neq \mu_2$ and the t in Eq. 6.4 obeys a central t distribution, the t_0 in Eq. 6.7 obeys a *noncentral* t distribution. More specifically, from Corollary 9 (Chap. 1 Sect. 1.3.1), we know that.[7]

$$t_0 \sim t'(\phi, \lambda_t), \qquad (6.8)$$

where

$$\lambda_t = \sqrt{n}\Delta_t, \quad \Delta_t = \frac{\mu_1 - \mu_2}{\sigma_d} = \frac{\mu_1 - \mu_2}{\sqrt{\sigma_1^2 + \sigma_2^2}}. \qquad (6.9)$$

Note that Δ_t is a form of population effect size expressed as a standardised mean difference, similar to Eq. 5.4 (Chap. 5 Sect. 5.2.1) but without the homoscedasticity assumption.

Going back to Table 6.1, *if H_0 is false*, Eq. 6.8 holds. Moreover, in a paired t-test, the probability of rejecting H_0 is given by Eq. 6.6 regardless of what distribution t_0 actually obeys. Hence, if H_0 is false, Eq. 6.6 is exactly $(1 - \beta)$, the statistical power: the probability of correctly concluding that the two population means are different. Henceforth, we consider Eq. 6.6 under Eq. 6.8 to evaluate statistical power.

Using Eq. 6.6 *as is* to compute statistical power is difficult because of the complexity of the noncentral t distribution. However, we can utilise Corollary 10 (Chap. 1 Sect. 1.3.1) to approximate it using a random variable $u \sim N(0, 1^2)$, with $\phi = n - 1$, $w = t_{inv}(\phi; \alpha)$, $\lambda_t = \sqrt{n}\Delta_t$:

$$1 - \beta \approx Pr\left\{u \leq \frac{-w(1 - 1/4\phi) - \lambda_t}{\sqrt{1 + w^2/2\phi}}\right\}$$

$$+ 1 - Pr\left\{u \leq \frac{w(1 - 1/4\phi) - \lambda_t}{\sqrt{1 + w^2/2\phi}}\right\}. \qquad (6.10)$$

Equation 6.10 is useful for estimating the statistical power with high accuracy when (α, Δ_t, n) are available; we shall come back to this formula later. However, since we want an estimate of n given $(\alpha, \beta, \Delta_t)$, here we follow an alternative path starting from Eq. 6.5, by first applying Corollary 11 (Chap. 1 Sect. 1.3.1), again using $u \sim N(0, 1^2)$ with $w = t_{inv}(\phi; \alpha)$:

$$1 - \beta \approx Pr\left\{u \leq \frac{-w - \lambda_t}{\sqrt{1 + w^2/2\phi}}\right\} + Pr\left\{u \geq \frac{w - \lambda_t}{\sqrt{1 + w^2/2\phi}}\right\}. \qquad (6.11)$$

If $\lambda_t > 0$, let us ignore the first term on the right side since it should be relatively small; similarly, if $\lambda_t < 0$, let us ignore the second term. Recall that we are

---

[7]In Corollary 9, let $\mu = \mu_1 - \mu_2, \sigma^2 = \sigma_1^2 + \sigma_2^2, \mu_0 = 0, \lambda = \lambda_t$.

considering the situation where H_0 is false and therefore $\lambda_t \neq 0$ (See Eq. 6.9). In either case, we obtain a further approximation:

$$1 - \beta \approx Pr\left\{ u \geq \frac{w - \lambda_t}{\sqrt{1 + w^2/2\phi}} \right\} . \tag{6.12}$$

Thus, since $Pr\{u \geq z_{inv}(P)\} = P$ constitutes the definition of $z_{inv}(P)$, the following should hold when $\lambda_t \neq 0$:

$$\frac{w - \lambda_t}{\sqrt{1 + w^2/2\phi}} \approx z_{inv}(1 - \beta) . \tag{6.13}$$

Meanwhile, Corollary 12 (Chap. 1 Sect. 1.3.1) gives us, for $w = t_{inv}(\phi; \alpha)$:

$$\frac{w}{\sqrt{1 + w^2/2\phi}} \approx z_{inv}(\alpha/2) . \tag{6.14}$$

Hence, from Eqs. 6.13 and 6.14, we obtain:

$$\lambda_t \approx (z_{inv}(\alpha/2) - z_{inv}(1 - \beta))\sqrt{1 + w^2/2\phi} . \tag{6.15}$$

By applying Corollary 14 (Chap. 1 Sect. 1.3.1) to the above, we have:

$$\lambda_t \approx \frac{z_{inv}(\alpha/2) - z_{inv}(1 - \beta)}{\sqrt{1 - z_{inv}(\alpha/2)^2/2\phi}} , \tag{6.16}$$

and hence:

$$\lambda_t^2 \approx \frac{(z_{inv}(\alpha/2) - z_{inv}(1 - \beta))^2}{1 - z_{inv}(\alpha/2)^2/2\phi} . \tag{6.17}$$

Finally, by replacing λ_t using Eq. 6.9 and letting $\phi = n - 1 \approx n$, we obtain:

$$n\Delta_t^2 \approx \frac{(z_{inv}(\alpha/2) - z_{inv}(1 - \beta))^2}{1 - z_{inv}(\alpha/2)^2/2n} , \tag{6.18}$$

and therefore

$$n \approx \left(\frac{z_{inv}(\alpha/2) - z_{inv}(1 - \beta)}{\Delta_t} \right)^2 + \frac{z_{inv}(\alpha/2)^2}{2} . \tag{6.19}$$

We have successfully expressed n in terms of α, β and the population effect size Δ_t.

Let $min\Delta_t$ be the smallest population effect size for which we want to guarantee $100(1 - \beta)\%$ power. By letting $\Delta_t = min\Delta_t$ in Eq. 6.19, we can estimate the minimally adequate topic set size. However, deriving Eq. 6.19 involved several approximations, so once we have obtained a topic set size from this formula, we should go back to Eq. 6.10 and check that the desired power is actually achieved. If not, n should be incremented until it achieves the desired power. For example, if we let $(\alpha, \beta, min\Delta_t) = (0.05, 0.20, 0.50)$, we initially obtain[8]:

$$n \approx \left(\frac{1.960 - (-0.842)}{0.50} \right)^2 + \frac{1.960^2}{2} = 33.3 \ . \tag{6.20}$$

Now, substitute $n = 33$ with $\Delta_t = min\Delta_t = 0.5$ to Eq. 6.10: since $w = t_{inv}(33 - 1; 0.05) = $ `T.INV.2T`$(0.05, 32) = 2.037$ (with Excel), the achieved power is:

$$1 - \beta \approx Pr\{u \leq -4.742\} + 1 - Pr\{u \leq -0.825\} = 0.795 \ , \tag{6.21}$$

which means that 80% power is not quite achieved. So we let $n = 34$, and the achieved power is computed similarly: $1 - \beta = 0.808$. Therefore, $n = 34$ is the topic set size we want.

The "`From effect size`" sheet of `samplesizeTTEST2` automates the above procedure. Table 6.2 shows the required topic set size for several combinations of $(\alpha, \beta, min\Delta_t)$.[9] It can be observed that:

- If we want a small α (i.e. Type I error probability), we need a large sample (compare top and bottom of Table 6.2);
- If we want a small β (i.e. Type II error probability), in other words, high statistical power, we need a large sample (compare the two columns of Table 6.2);
- If we want to guarantee a given combination of error probabilities (α, β) even for a small effect size, we need a large sample.

Recall from Eq. 6.9 that Δ_t is a population effect size expressed as a *standardised* mean difference. If we have an estimate of the population variance $\sigma_d^2 = \sigma_1^2 + \sigma_2^2$ for a particular evaluation measure, we can impose our statistical requirement in terms of an *raw* mean difference with that evaluation measure instead of $min\Delta$. That is, we can require $100(1 - \beta)\%$ statistical power whenever $|\mu_1 - \mu_2| \geq minD_t$ for a particular evaluation measure. Given such a requirement and a particular variance estimate $\hat{\sigma}_d^2$, we can convert the above minimum detectable difference into a $min\Delta_t$ using the definition of Δ_t in Eq. 6.9:

$$min\Delta_t = \frac{minD_t}{\sqrt{\hat{\sigma}_d^2}} \ . \tag{6.22}$$

---

[8]Recall that with Microsoft Excel, $z_{inv}(P)$ can be obtained as `NORM.S.INV`$(1 - P)$.

[9]This table corrects a typo in Table 1 of Sakai [22] for $(\alpha, \beta, min\Delta_t) = (0.05, 0.20, 1.0)$, and provides the sample sizes for $min\Delta_t = 1.5, 2.0$ in addition.

Table 6.2 Topic set sizes obtained with `samplesizeTTEST2` (sheet: `From effect size`) for $(\alpha, \beta, min\Delta_t)$

α	$min\Delta_t$	$\beta = 0.10$	$\beta = 0.20$
0.01	0.1	1492	1172
	0.2	376	296
	0.5	63	51
	1.0	19	16
	1.5	11	9
	2.0	8	7
0.05	0.1	1053	787
	0.2	265	199
	0.5	44	34
	1.0	13	10
	1.5	7	6
	2.0	5	5

This is why the "`From absolute diff`" sheet of `samplesizeTTEST2` requires $(\alpha, \beta, minD_t, \hat{\sigma}_d^2)$ instead of $(\alpha, \beta, min\Delta_t)$.

In Sect. 6.8, we shall discuss topic set size design for the paired t-test with $(\alpha, \beta, minD_t, \hat{\sigma}_d^2)$ using real data from IR.

6.3 Topic Set Size Design with the Two-Sample t-Test

6.3.1 How to Use the Two-Sample t-Test-Based Tool

`samplesize2SAMPLET` requires the following as input:

α Type I error probability.

β Type II error probability. That is, you want $100(1 - \beta)\%$ statistical power for any two-sample t-test (see below).

$minD_t$ Minimum detectable difference. That is, whenever the true difference between two systems (in terms of a particular evaluation measure) is $minD_t$ or larger, you want to guarantee $100(1 - \beta)\%$ statistical power.

$\hat{\sigma}^2$ An estimate of the common score variance in terms of the above evaluation measure, under the homoscedasticity assumption. How to obtain such an estimate will be discussed in Sect. 6.7.

By entering the above values into the "`From absolute diff`" sheet of the Excel file, the topic set size required for a balanced design, i.e. $n = n_1 = n_2$, can easily be obtained. The other sheet, "`From effect size`", will be discussed in Sect. 6.3.2.

Suppose you are interested in the statistical power for comparing any system pair with a two-sample t-test at $\alpha = 0.05$ with an evaluation measure whose estimated population variance is $\hat{\sigma}^2 = 0.1$. If you want to guarantee 80% statistical power whenever the true difference between any two systems is 0.1 or larger, enter

	A	B	C	D	E	F	G
1	alpha	beta	minDelta_t	minD_t	sigma^2		
2	0.05	0.2	0.3162	0.1	0.1		
3	zalpha/2	z1−beta	n approx	n	large enough?	w	u <= ? (−w)
4	1.959964	−0.841621	157.9	161	TRUE	1.967405	−4.78866119
5	README		n recommended	160	TRUE	1.9674519	−4.7798088
6			158	159	TRUE	1.9674995	−4.77092852
7			large enough?	158	TRUE	1.9675477	−4.76202009
8			TRUE	157	FALSE	1.9675965	−4.75308324
9				156	FALSE	1.9676459	−4.7441177
10				155	FALSE	1.967696	−4.73512319
11				154	FALSE	1.9677467	−4.72609942
12				153	FALSE	1.9677981	−4.71704611
13				152	FALSE	1.9678502	−4.70796296

Fig. 6.2 How to use `samplesize2SAMPLET`: an example (sheet: `From absolute diff`)

$(\alpha, \beta, minD_t, \hat{\sigma}^2) = (0.05, 0.20, 0.10, 0.10)$ into the "`From absolute diff`" sheet. Figure 6.2 shows what happens: it can be observed that the required topic set size is 158: this is the minimum topic set size for a balanced design that satisfies the required statistical power under the specified condition.[10] That is, each of the two topic sets should contain 158 topics.

6.3.2   How the Two-Sample t-Test-Based Topic Set Size Design Works

How the power-based topic set size design tool `samplesize2SAMPLET` for the two-sample t-test works is very similar to how `samplesizeTTEST2` for the paired t-test works (See Sect. 6.2.2). Therefore, I shall only highlight the main differences in this section.

First, recall that the test statistic for the two-sample t-test is given by (Chap. 2 Sect. 2.3):

$$t_0 = \frac{\bar{x}_{1\bullet} - \bar{x}_{2\bullet}}{\sqrt{V_p(1/n_1 + 1/n_2)}} \, . \tag{6.23}$$

If $H_0 : \mu_1 = \mu_2$ is true, t_0 obeys $t(\phi)$ where $\phi = n_1 + n_2 - 2$. On the other hand, if H_0 is false, t_0 obeys a noncentral t distribution $t'(\phi, \lambda_t)$, where

$$\lambda_t = \sqrt{\frac{n_1 n_2}{n_1 + n_2}} \Delta_t \, , \quad \Delta_t = \frac{\mu_1 - \mu_2}{\sigma} \, , \tag{6.24}$$

---

[10]The achieved power is computed in Column K, although not shown in Fig. 6.2.

and σ is the common population variance. Compare this with Eq. 6.9 that we discussed for the paired data case.

Henceforth we consider a balanced design $n = n_1 + n_2$, so that:

$$\phi = 2n - 2 , \quad \lambda = \sqrt{\frac{n}{2}} \Delta_t . \tag{6.25}$$

The statistical power for the two-sample t-test can be computed as before using Eq. 6.10 but with the above ϕ and λ. This is the formula we should go back to to check the achieved power.

As for the initial topic set size estimate, let us start from Eq. 6.16. By substituting the λ_t with the above and letting $\phi = 2n - 2 \approx 2n$, we obtain:

$$\frac{n\Delta_t^2}{2} \approx \frac{(z_{inv}(\alpha/2) - z_{inv}(1 - \beta))^2}{1 - z_{inv}(\alpha/2)^2/4n} , \tag{6.26}$$

and therefore

$$n \approx 2 \left(\frac{z_{inv}(\alpha/2) - z_{inv}(1 - \beta)}{\Delta_t} \right)^2 + \frac{z_{inv}(\alpha/2)^2}{4} . \tag{6.27}$$

As with the paired data case, let $min\Delta_t$ be the smallest population effect size for which we want to guarantee $100(1 - \beta)\%$ power. We substitute $\Delta_t = min\Delta_t$ to Eq. 6.27 to obtain an initial topic set size estimate, check the achieved power with Eq. 6.10, increment n if necessary, and so on. Table 6.3 shows the results for several combinations of $(\alpha, \beta, min\Delta_t)$, using the "From effect size" sheet of samplesize2SAMPLET, which automates the above process. It can be observed that the topic set sizes in this table are much larger compared to those shown for the paired data case in Fig. 6.2. However, these tables are not directly comparable, since while the $min\Delta_t$ for paired data case measures effects based on the standard deviation of the score *differences* given by $\sigma_d = \sqrt{\sigma_1^2 + \sigma_2^2}$ (See Eq. 6.9), the $min\Delta_t$ for unpaired data measures effects based on the standard deviation σ of the scores (See Eq. 6.24).

Again, for a particular evaluation measure whose variance estimate is $\hat{\sigma}^2$, we can specify a minimum detectable difference $minD$ instead of $min\Delta_t$ and require $100(1 - \beta)\%$ power whenever $|\mu_1 - \mu_2| \geq minD_t$. The "From absolute diff" sheet of samplesize2SAMPLET requires $(\alpha, \beta, minD_t, \hat{\sigma}^2)$ and uses the definition of Δ_t in Eq. 6.24 to convert the minimum detectable difference into $min\Delta_t$ for topic set size design as follows:

$$min\Delta_t = \frac{minD_t}{\sqrt{\hat{\sigma}^2}} . \tag{6.28}$$

Table 6.3 Topic set sizes obtained with `samplesize2SAMPLET` (sheet: `From effect size`) for $(\alpha, \beta, min\Delta_t)$

α	$min\Delta_t$	$\beta = 0.10$	$\beta = 0.20$
0.01	0.1	2978	2338
	0.2	746	586
	0.5	121	96
	1.0	32	26
	1.5	15	13
	2.0	10	8
0.05	0.1	2103	1571
	0.2	527	394
	0.5	86	64
	1.0	23	17
	1.5	11	9
	2.0	7	6

In Sect. 6.8, we shall discuss topic set size design for the two-sample t-test with $(\alpha, \beta, minD_t, \hat{\sigma}_d^2)$ using real data from IR.

6.4 Topic Set Size Design with One-Way ANOVA

`samplesizeANOVA2` is a tool for computing the required number of topics in order to achieve a specified statistical power in one-way ANOVA for comparing $m(\geq 2)$ systems.

6.4.1 How to Use the ANOVA-Based Tool

`samplesizeANOVA2` requires the following as input:

α Type I error probability.

β Type II error probability. That is, you want $100(1 - \beta)\%$ statistical power for any one-way ANOVA test (see below).

m Number of systems that will be compared based on one-way ANOVA.

$minD$ *Minimum detectable range*. That is, whenever the true difference between the best and the worst among the m systems (in terms of a particular evaluation measure) is *minD* or larger, you want to guarantee $100(1 - \beta)\%$ statistical power.

$\hat{\sigma}^2$ Variance estimate for the scores in terms of the above evaluation measure. How to obtain such an estimate will be discussed in Sect. 6.7.

	A	B	C	D	E	F	G
1	minD	sigma^2	m	lambda	n approx	n recommended	large enough?
2	0.1	0.1	10	15.612	312.2	312	TRUE
3	m	n	large enough	phiE	w	cA	phiA*
4	10	412	TRUE	4110	1.8821558	1.695945946	17.45338645
5	10	411	TRUE	4100	1.8821614	1.695431472	17.42919162
6	10	410	TRUE	4090	1.8821669	1.694915254	17.405
7	10	409	TRUE	4080	1.8821725	1.694397284	17.38081162
8	10	408	TRUE	4070	1.8821781	1.693877551	17.35662651
9	10	407	TRUE	4060	1.8821838	1.693356048	17.33244467
10	10	406	TRUE	4050	1.8821895	1.692832765	17.30826613

Fig. 6.3 How to use `samplesizeANOVA2`: an example (sheet: `alpha=.05, beta=.20`)

In fact, this tool only accommodates combinations of (α, β) for $\alpha = 0.01, 0.05$ and $\beta = 0.10, 0.20, 0.30, 0.40, 0.50$.[11] This is not a problem if Cohen's five-eighty convention ($\alpha = 0.05$, $\beta = 0.20$) is to be followed. By entering $(m, minD, \hat{\sigma}^2)$ into one of the Excel sheets that represents a particular combination of α and β, the topic set size required can easily be obtained.

Suppose you are interested in the statistical power for comparing any $m = 10$ systems with one-way ANOVA at $\alpha = 0.05$ with an evaluation measure whose estimated population variance is $\hat{\sigma}^2 = 0.1$. If you want to guarantee 80% statistical power whenever the true difference between the best and the worst systems is 0.1 or larger, enter $(m, minD, \hat{\sigma}^2) = (10, 0.10, 0.10)$ into the "`alpha=.05, beta=.20`" sheet. Figure 6.3 shows what happens: it can be observed that the required topic set size is 312: this is the minimum topic set size that satisfies the required statistical power under the specified condition.[12]

6.4.2   How the ANOVA-Based Topic Set Size Design Works

This section explains how `samplesizeANOVA2` obtains the required sample size for one-way ANOVA, given $(\alpha, \beta, m, minD, \hat{\sigma}^2)$. First, let us recall what we do when we conduct one-way ANOVA for comparing m systems with equal group sizes (Chap. 3 Sect. 3.1.1): the appropriate choice for the common group size is what we want here. As was mentioned earlier, sample sizes obtained using this tool can be regarded as pessimistic estimates for cases where two-way ANOVA without replication is more appropriate. That is, we allow ourselves to err on the side of oversampling.

We start with the homoscedasticity and the normality assumptions:

[11] An earlier version of this tool, `samplesizeANOVA`, accommodates only $\alpha = 0.01, 0.05$ and $\beta = 0.10, 0.20$ [22].
[12] The achieved power is computed in Column I, although not shown in Fig. 6.3.

$$x_{ij} \sim N(\mu_i, \sigma^2) \,, \tag{6.29}$$

and define the population grand mean μ and the i-th system effect as:

$$\mu = \frac{\sum_{i=1}^{m} \mu_i}{m}, \quad a_i = \mu_i - \mu \,. \tag{6.30}$$

Recall that the test statistic for one-way ANOVA is given by:

$$F_0 = \frac{V_A}{V_{E1}} \,, \quad V_A = \frac{S_A}{\phi_A} \,, \quad V_{E1} = \frac{S_{E1}}{\phi_{E1}} \,. \tag{6.31}$$

The null hypothesis is given by $H_0 : a_1 = a_2 = \cdots = a_m = 0$; we reject H_0 iff $F_0 \geq F_{inv}(\phi_A, \phi_{E1}; \alpha)$. Thus, the probability of rejecting H_0 is given by:

$$Pr\{F_0 \geq F_{inv}(\phi_A, \phi_{E1}; \alpha)\} = 1 - Pr\{F_0 \leq F_{inv}(\phi_A; \phi_{E1}, \alpha)\} \,. \tag{6.32}$$

The following argument is similar to the one we had for the paired t-test in Sect. 6.2.2. If H_0 is true, then $F_0 \sim F(\phi_A, \phi_{E1})$ and Eq. 6.32 are exactly α: that is how $F_{inv}(\phi_A; \phi_{E1}, \alpha)$ is defined. On the other hand, if H_0 is false, Eq. 6.32 represents $(1 - \beta)$, the statistical power, and it is known that $F_0 \sim F'(\phi_A, \phi_{E1}; \lambda)$, i.e. a noncentral F distribution (Chap. 3 Sect. 1.3.3), where the noncentrality parameter is given by [16]:

$$\lambda = n\Delta \,, \quad \Delta = \frac{\sum_{i=1}^{m} a_i^2}{\sigma^2} = \frac{\sum_{i=1}^{m}(\mu_i - \mu)^2}{\sigma^2} \,. \tag{6.33}$$

Note that Δ measures the sum of squared system effects in variance units.

To compute the power based on Eq. 6.32, we utilise Theorem 13 (Chap. 1 Sect. 1.3.3):

$$1 - \beta \approx 1 - Pr\{u \leq \frac{\sqrt{\frac{w}{\phi_{E1}}}\sqrt{2\phi_{E1} - 1} - \sqrt{\frac{c_A}{\phi_A}}\sqrt{2\phi_A^* - 1}}{\sqrt{\frac{c_A}{\phi_A} + \frac{w}{\phi_{E1}}}}\} \tag{6.34}$$

where $u \sim N(0, 1^2)$, $w = F_{inv}(\phi_A; \phi_{E1}, \alpha)$ and

$$c_A = \frac{\phi_A + 2\lambda}{\phi_A + \lambda} = \frac{\phi_A + 2n\Delta}{\phi_A + n\Delta} \,, \tag{6.35}$$

$$\phi_A^* = \frac{(\phi_A + \lambda)^2}{\phi_A + 2\lambda} = \frac{(\phi_A + n\Delta)^2}{\phi_A + 2n\Delta} \,. \tag{6.36}$$

Recall that, in the case of the topic set size design based on t-tests, we specified $minD_t$, the minimum detectable difference for ensuring $100(1 - \beta)\%$ power.

However, in the case of ANOVA, we have $m(\geq 2)$ systems, and *all* of these systems affect the effect size Δ shown in Eq. 6.33. Hence, to clarify the condition for ensuring $100(1 - \beta)\%$ power, let us consider the *range* of the m population means, $D = \max_i a_i - \min_i a_i$, i.e. the true difference between the best and the worst among the m systems. We require $100(1 - \beta)\%$ power whenever $D \geq minD$, where $minD$ is the aforementioned minimum detectable *range*.

Now, let us define $min\Delta$ as follows:

$$min\Delta = \frac{minD^2}{2\sigma^2} \, . \tag{6.37}$$

Since $\sum_{i=1}^{m} a_i^2 \geq \frac{D^2}{2}$ holds,[13] we can see that:

$$\Delta = \frac{\sum_{i=1}^{m} a_i^2}{\sigma^2} \geq \frac{D^2}{2\sigma^2} \geq \frac{minD^2}{2\sigma^2} = min\Delta \, . \tag{6.38}$$

That is, while specifying $minD$ does not uniquely determine Δ (as it depends on systems other than the best and the worst ones), $min\Delta$ as defined in Eq. 6.37 is a lowerbound for the actual Δ. Hence, by plugging in Eqs. 6.37 to 6.34, we can obtain the worst-case estimate of the statistical power. Similarly, we should be able to obtain a worst-case estimate of the topic set size from Eq. 6.33:

$$n \approx \frac{\lambda}{min\Delta} = \frac{2\sigma^2\lambda}{minD^2} \, . \tag{6.39}$$

In Eq. 6.39, $minD$ is the minimum detectable range that we specify as input; the common population variance σ^2 can be estimated from data as we shall discuss in Sect. 6.7. Therefore what we have left to compute for estimating the topic set size n is λ. To do this, we look at Eq. 6.32, let $\phi_{E1} = \infty$ and apply Corollary 16 (Chap. 1 Sect. 1.3.3). That is, we approximate the noncentral F distribution $F'(\phi_A, \phi_E, \lambda)$ with a noncentral χ^2 distribution $\chi'^2(\phi_A, \lambda)$. For $\chi'^2(\phi_A, \lambda)$, Nagata [16] provides linear approximations of λ for $\alpha = 0.01, 0.05$ and $\beta = 0.10, 0.20, 0.30, 0.40, 0.50$ as shown in Table 6.4; this is why samplesizeANOVA2 has ten different sheets. Thus, we can now obtain an initial estimate of the topic set size using Eq. 6.39.

Suppose that we require $(\alpha, \beta, minD, m) = (0.05, 0.20, 0.10, 10)$ and that we obtained $\hat{\sigma}^2 = 0.10$ from past data using a method discussed in Sect. 6.8. Then $min\Delta \approx \frac{minD^2}{2\hat{\sigma}^2} = 0.10^2/(2 * 0.10) = 0.05$. Since $\phi_A = m - 1 = 10 - 1 = 9$, $\lambda = 4.860 + 3.584\sqrt{9} = 15.612$, and hence $n = \lambda/min\Delta \approx 312.2$. Suppose we let $n = 312$. Then $\phi_{E1} = m(n - 1) = 10(312 - 1) = 3110$, $w = F_{inv}(9, 3110; 0.05) =$ F.INV.RT$(0.05, 9, 3110) = 1.883$ (with Excel). From Eqs. 6.35 and 6.36, $c_A = 1.634$, $\phi_A^* = 15.054$; hence from Eq. 6.34, the achieved

---

[13]Let $A = \max_i a_i$ and $a = \min_i a_i$. Then $D^2/2 = (A^2 + a^2 - 2Aa)/2 \leq A^2 + a^2 \leq \sum_{i=1}^{m} a_i^2$. The equality holds when $A = D/2, a = -D/2$ and $a_i = 0$ for all other systems.

Table 6.4 Linear approximations of λ, the noncentrality parameter of a noncentral χ^2 distribution $\chi'^2(\phi_A, \lambda)$ [16]

β	$\alpha = 0.01$	$\alpha = 0.05$
0.1	$10.439 + 5.213\sqrt{\phi_A}$	$7.049 + 4.244\sqrt{\phi_A}$
0.2	$7.736 + 4.551\sqrt{\phi_A}$	$4.860 + 3.584\sqrt{\phi_A}$
0.3	$6.030 + 4.077\sqrt{\phi_A}$	$3.522 + 3.112\sqrt{\phi_A}$
0.4	$4.733 + 3.675\sqrt{\phi_A}$	$2.543 + 2.712\sqrt{\phi_A}$
0.5	$3.653 + 3.302\sqrt{\phi_A}$	$1.765 + 2.341\sqrt{\phi_A}$

power is $1 - Pr\{u \leq -0.843\} = 0.800$, which is what we want. However, is this topic set size minimal? If we try $n = 311$, it can be verified similarly that the achieved power is only $1 - Pr\{u \leq -0.837\} = 0.799$, and therefore that $n = 312$ is indeed minimal. samplesizeANOVA2 simply automates the above procedure; Fig. 6.3 in fact shows the above exact example using the "alpha=.05, beta=0.20" sheet.

In the above example, our initial estimate ($n = 312$) happened to be spot on: it was the minimal topic set size required to achieve the desired statistical power. However, as we shall demonstrate in Sect. 6.8.1.3, the initial estimate can be quite rough, and the final topic set size may be substantially different from the initial estimate. Hence, checking the achieved power using Eq. 6.34 is in fact important. samplesizeANOVA2 automatically tries a range of numbers larger and smaller than the initial estimate and finds the minimal n that satisfies Eq. 6.34.

6.5 Topic Set Size Design with Confidence Intervals for Paired Data

6.5.1 How to Use the Paired-Data CI-Based Tool

samplesizeCI2 is a tool for computing the required number of topics based on a desired cap on the expected width of the CI for the difference between any pair of systems for paired data.

samplesizeCI2 requires the following as input:

α Type I error probability. That is, you want to consider $100(1 - \alpha)\%$ CIs.

δ A cap on the expected width of the CI for the difference between any two systems, in terms of a particular evaluation measure. That is, you want the expected CI width for the difference between any system pair to be δ or smaller.

$\hat{\sigma}_d^2$ Variance estimate for the score differences in terms of the above evaluation measure. How to obtain such an estimate will be discussed in Sect. 6.7.

By entering the above values into the "approximation" sheet of the Excel file, the topic set size required can easily be obtained. Suppose you want the expected width of a 95% CI for the difference between any system pair to be no larger than

	A	B	C		D	E	F	G
1	alpha	delta (<=2ME)	sigma_d^2		README			
2	0.05	0.1	0.2					
3	zalpha/2		n approx		n	RHS	large enough?	LHS
4	0.05	1.9600		307.3	317	0.111803399	TRUE	0.1104184
5	0.05		n recommended		316	0.111803399	TRUE	0.110594
6	0.05			310	315	0.111803399	TRUE	0.1107705
7	0.05		large enough?		314	0.111803399	TRUE	0.1109478
8	0.05		TRUE		313	0.111803399	TRUE	0.111126
9	0.05				312	0.111803399	TRUE	0.1113051
10	0.05				311	0.111803399	TRUE	0.111485
11	0.05				310	0.111803399	TRUE	0.1116658
12	0.05				309	0.111803399	FALSE	0.1118475
13	0.05				308	0.111803399	FALSE	0.11203

Fig. 6.4 How to use `samplesizeCI2`: an example (sheet: `approximation`)

0.1 for an evaluation measure whose estimated population variance for the score differences is $\hat{\sigma}_d^2 = 0.2$. Enter $(\alpha, \delta, \hat{\sigma}_d^2) = (0.05, 0.10, 0.20)$. Figure 6.4 shows what happens: it can be observed that the required topic set size is 310: this is the minimum topic set size required to satisfy the specified condition.

6.5.2   How the Paired-Data CI-Based Tool Works

This section explains how `samplesizeCI2` obtains the required sample size given $(\alpha, \beta, \delta, \hat{\sigma}_d^2)$.

Recall how we constructed the MOE for a $100(1 - \alpha)\%$ CI for the difference between two system means in Chap. 2 Sect. 2.8; the basic assumptions are the same as those used for the paired t-test (see Sect. 6.2.2). That is, we assume that:

$$d_j \sim N(\mu_1 - \mu_2, \sigma_d^2), \quad \sigma_d^2 = \sigma_1^2 + \sigma_2^2, \tag{6.40}$$

and the MOE is given by:

$$MOE = t_{inv}(\phi; \alpha)\sqrt{V_d/n}, \quad \phi = n - 1. \tag{6.41}$$

Since the CI is given by $[\bar{d} - MOE, \bar{d} + MOE]$, the width of the CI is $2MOE$. Here, we consider the approach of determining the topic set size required to ensure $2MOE \leq \delta$ for the difference between any system pair. Recall, for example, that a wide 95%CI that includes zero means that the difference is not statistically significant at $\alpha = 0.05$; we generally want tight CIs.

Equation 6.41 contains a random variable, namely, V_d. Therefore, the actual requirement we impose on the CI width is in terms of the *expectation* of $2MOE$:

$$E(2MOE) = 2t_{inv}(\phi; \alpha)\frac{E(\sqrt{V_d})}{\sqrt{n}} \leq \delta. \tag{6.42}$$

Now, it is known that:

$$E(\sqrt{V_d}) = c^* \sigma_d , \quad c^* = \frac{\sqrt{2}\Gamma((\phi + 1)/2)}{\sqrt{\phi}\Gamma(\phi/2)} , \tag{6.43}$$

where $\Gamma(\bullet)$ is the gamma function (Chap. 1 Sect. 1.2.3).[14] Note that, while V_d is an unbiased estimator of σ_d^2, i.e. $E(V_d) = \sigma_d^2$, Eq. 6.43 states that $\sqrt{V_d}$ is not an unbiased estimator of the population standard deviation σ_d.

By substituting Eq. 6.43 to Eq. 6.42, the requirement can be rewritten as:

$$2t_{inv}(\phi; \alpha)\frac{c^* \sigma_d}{\sqrt{n}} \leq \delta . \tag{6.44}$$

Moreover, since computing c^* is computationally highly expensive, we leverage a highly accurate approximation used in Corollary 10 (Chap. 1 Sect. 1.3.1), namely, $c^* \approx 1 - 1/4\phi$, to rewrite the above:

$$\frac{t_{inv}(\phi; \alpha)(1 - 1/4\phi)}{\sqrt{n}} \leq \frac{\delta}{2\sigma_d} , \quad \phi = n - 1 . \tag{6.45}$$

In order to find the smallest n that satisfies Eq. 6.45, let us go back to the original requirement given by Eq. 6.42, and consider an "easy" case where the population variance σ_d^2 of the score differences is *known*. That is, we tentatively consider a standard normal distribution rather than a t-distribution.[15] By replacing the two-sided t-value $t_{inv}(\phi; \alpha)$ with a one-sided z-value and the expectation of the sample standard deviation $E(\sqrt{V_d})$ with σ_d, we obtain:

$$2z_{inv}(\alpha/2)\frac{\sigma_d}{\sqrt{n}} \leq \delta , \tag{6.46}$$

or

$$n \geq \frac{4z_{inv}(\alpha/2)^2 \sigma_d^2}{\delta^2} . \tag{6.47}$$

Thus, provided we have a variance estimate $\hat{\sigma}_d^2$, we can first obtain the smallest integer that satisfies Eq. 6.47 and then check whether the original requirement given

---

[14]Let χ^2 be a random variable that obeys $\chi^2(\phi)$. Then c^* represents the population mean of the random variable $\sqrt{\chi^2/\phi}$. That is, $E(\sqrt{\chi^2/\phi}) = c^*$. This is the same c^* used in Theorem 11 (Chap. 1 Sect. 1.3.1).

[15]Recall Corollary 5 (Chap. 1 Sect. 1.2.4): if $u = \frac{\bar{x}-\mu}{\sqrt{\sigma^2/n}} \sim N(0, 1^2)$, then $t = \frac{\bar{x}-\mu}{\sqrt{V/n}} \sim t(n - 1)$ where $E(V) = \sigma^2$. That is, a t-distribution is like the standard normal distribution, except that there is an uncertainty about the estimator of σ^2, whose accuracy increases with n.

by Eq. 6.45 is satisfied. If not, the topic set size is incremented one by one until the condition is met.

Suppose that we let $(\alpha, \delta, \hat{\sigma}_d^2) = (0.05, 0.10, 0.20)$. From Eq. 6.47, $n \geq 4 *$ NORM.S.INV$(1 - 0.025)^2 * 0.20/0.10^2 = 307.3$, but Eq. 6.45 is not satisfied by $n = 308$ and $n = 309$. Finally, with $n = 310$, the left side of Eq. 6.45 is smaller than the right side. This is exactly what is happening in Fig. 6.4.

An early version of samplesizeCI2, called samplesizeCI [22], involved a direct computation of the above c^* instead of leveraging Eq. 6.45. For this reason, it could not handle large topic set sizes: GAMMA(172) is greater than 10^{307} and cannot be computed by Excel. This older method is preserved in the "using Gamma" sheet of samplesizeCI2. However, I recommend the reader to use the "approximation" sheet as the approximation $c^* \approx 1 - 1/4\phi$ is highly accurate except when $\phi \approx 1$.

6.6   Topic Set Size Design with Confidence Intervals for Unpaired Data

6.6.1   How to Use the Two-Sample CI-Based Tool

samplesize2SAMPLECI is a tool for computing the required number of topics based on a desired cap on the expected width of the CI for the difference between any pair of systems for unpaired data.

samplesize2SAMPLECI requires the following as input:

α Type I error probability. That is, you want to consider $100(1 - \alpha)\%$ CIs.

δ A cap on the expected width of the CI for the difference between any two systems, in terms of a particular evaluation measure. That is, you want the expected CI width for the difference between any system pair to be δ or smaller.

$\hat{\sigma}^2$ An estimate of the common score variance in terms of the above evaluation measure, under the homoscedasticity assumption. How to obtain such an estimate will be discussed in Sect. 6.7.

By entering the above values into the "approximation" sheet of the Excel file, the topic set size required can easily be obtained. Suppose you want the expected width of a 95% CI for the difference between any system pair to be no larger than 0.1 for an evaluation measure whose estimated population variance is $\hat{\sigma}^2 = 0.1$. Enter $(\alpha, \delta, \hat{\sigma}^2) = (0.05, 0.10, 0.10)$. Figure 6.5 shows what happens: it can be observed that the required topic set size is 309: this is the minimum topic set size required to satisfy the specified condition.

	A	B	C	D	E	F	G
1	alpha	delta (<=2ME)	sigma^2	README			
2	0.05	0.1	0.1				
3	zalpha/2	n approx		n	RHS	large enough?	LHS
4	0.05	1.9600	307.3	317	0.1118034	TRUE	0.1102501
5	0.05	n recommended		316	0.1118034	TRUE	0.110425
6	0.05		309	315	0.1118034	TRUE	0.1106007
7	0.05	large enough?		314	0.1118034	TRUE	0.1107772
8	0.05	TRUE		313	0.1118034	TRUE	0.1109545
9	0.05			312	0.1118034	TRUE	0.1111328
10	0.05			311	0.1118034	TRUE	0.1113118
11	0.05			310	0.1118034	TRUE	0.1114918
12	0.05			309	0.1118034	TRUE	0.1116726
13	0.05			308	0.1118034	FALSE	0.1118543

Fig. 6.5 How to use `samplesize2SAMPLECI`: an example (sheet: `approximation`)

6.6.2 How the Two-Sample CI-Based Tool Works

How the CI-based topic set size design tool `samplesize2SAMPLECI` for unpaired data works is very similar to how `samplesizeTTEST2` for paired data works (see Sect. 6.5.2). Therefore, I shall only highlight the main differences in this section.

For the unpaired (i.e. two-sample) case, we consider a balanced design, i.e. $n = n_1 = n_2$ with $\phi = n_1 + n_2 - 2 = 2n - 2$. Hence the CI width requirement analogous to Eq. 6.44 is given by:

$$2t_{inv}(\phi; \alpha)E(\sqrt{V_p})\sqrt{\frac{1}{n_1} + \frac{1}{n_2}} = 2t_{inv}(\phi; \alpha)c^*\sigma\sqrt{\frac{2}{n}} \leq \delta . \tag{6.48}$$

Again, by applying $c^* \approx 1 - 1/4\phi$, we obtain the following requirement, which is analogous to Eq. 6.45:

$$\frac{t_{inv}(\phi; \alpha)(1 - 1/4\phi)}{\sqrt{n}} \leq \frac{\delta}{2\sqrt{2}\sigma} , \quad \phi = 2n - 2 . \tag{6.49}$$

As for the initial topic set size estimate that is analogous to Eq. 6.47, consider a standard normal distribution version of Eq. 6.48:

$$2z_{inv}(\alpha/2)\sigma\sqrt{\frac{2}{n}} \leq \delta , \tag{6.50}$$

or

$$n \geq \frac{8z_{inv}(\alpha/2)^2\sigma^2}{\delta^2} . \tag{6.51}$$

As before, we start with the smallest integer that satisfies Eq. 6.51 and increment it until it satisfies Eq. 6.49.

6.7   Estimating Population Variances

As was mentioned in Sect. 6.1, `samplesize2SAMPLET`, `samplesizeANOVA2`, and `samplesize2SAMPLECI` are for unpaired data, and they require an estimate of the common population variance σ^2, for a given evaluation measure, as input. On the other hand, `samplesizeTTEST2` and `samplesizeCI2` are for paired data, and they require an estimate of the population variance σ_d^2 for the score *differences* as input.

In order to obtain a estimate $\hat{\sigma}^2$, one needs to have at least one topic-by-run score matrix for the evaluation measure of your choice. Sakai and Shang [25] report that if we have runs from a few different teams for about 25 topics, reasonable estimates can be obtained. Thus, even if one is planning to build a new test collection for a new task, it is recommended that the evaluation measure be tested on an existing data set with several different systems. Starting with a small data set and hence an initial estimate of the variance is probably better than blindly creating a new test collection with an arbitrary choice for the number of topics. Once the test collection is completed, a larger and therefore more reliable topic-by-run matrix can be obtained, from which a more reliable variance estimate can be obtained for further improving the topic set size design for the next round of the same task. Thus, test collections can evolve.

Let C denote a pilot topic-by-run matrix with n_C topics and m_C systems. Under the homoscedasticity assumption, the residual mean squares from one-way ANOVA (Chap. 3 Sect. 3.1.1) or two-way ANOVA without replication (Chap. 3 Sect. 3.2) provide unbiased estimates of the population variance σ^2:

$$\hat{\sigma}_C^2 = V_{E1} = \frac{\sum_{i=1}^{m_C}\sum_{j=1}^{n_C}(x_{ij}-\bar{x}_{i\bullet})^2}{m_C(n_C-1)}, \tag{6.52}$$

$$\hat{\sigma}_C^2 = V_{E2} = \frac{\sum_{i=1}^{m_C}\sum_{j=1}^{n_C}(x_{ij}-\bar{x}_{i\bullet}-\bar{x}_{\bullet j}+\bar{x})^2}{(m_C-1)(n_C-1)} \tag{6.53}$$

where

$$\bar{x}_{i\bullet}=\frac{1}{n_C}\sum_{j=1}^{n_C}x_{ij},\quad \bar{x}_{\bullet j}=\frac{1}{m_C}\sum_{i=1}^{m_C}x_{ij},\quad \bar{x}=\frac{1}{m_C n_C}\sum_{i=1}^{m_C}\sum_{j=1}^{n_C}x_{ij}. \tag{6.54}$$

Recall that while V_{E1} is the mean squares obtained after removing the system sum of squares from the total sum of squares, V_{E2} is the one obtained after further removing the topic sum of squares. Hence, if one requires a tight variance estimate V_{E2} is recommended; however, if one is happy to err on the side of oversampling, V_{E1} can be used.

If there are multiple topic-by-run matrices for the same evaluation measures, the variances obtained above may be pooled for a better estimate:

$$\hat{\sigma}^2 = \frac{\sum_C (n_C - 1)\hat{\sigma}_C^2}{\sum_C (n_C - 1)} . \tag{6.55}$$

For example, if a topic-by-run matrix is obtained from each round of the same task, the variance estimates can be pooled for the topic set size design of the next round.

The above methods can be used for the unpaired-data topic set size design tools `samplesize2SAMPLET`, `samplesizeANOVA2`, and `samplesize2SAMPLECI`. As for the paired-data tools `samplesizeTTEST2` and `samplesizeCI2` that require an estimate of $\sigma_d^2 = \sigma_1^2 + \sigma_2^2$, a simple approach would be to just let $\hat{\sigma}_d^2 = 2\hat{\sigma}$, by accepting the homoscedasticity assumption *and* assuming that the scores from the two systems are *uncorrelated*.[16] While this approach overestimates σ_d^2 as we shall demonstrate in Sect. 6.8, obtaining tighter estimates of σ_d^2 is outside the scope of this book.

Table 6.5 duplicates some variance estimates from Sakai [22] (but rounded to three decimal places), where σ_C^2 were obtained using Eq. 6.52 and the pooled variances were obtained using Eq. 6.55. The data are from the TREC 2003–2004 robust retrieval tracks [29, 30] and the TREC 2011–2012 web track ad hoc tasks [9, 10]. The second column shows the *measurement depth* or *md*, which means that only the top *md* documents of each run were evaluated. The third and fourth columns show the sizes of the topic-by-run matrices. The remaining four columns show the variance estimates for the following evaluation measures, respectively:

AP   *Average Precision*: a binary-relevance measure that was widely used in early TREC tracks [3].

Q   *Q-measure* [19]. Similar to AP but can handle graded relevance. Widely used in NTCIR (NII Testbeds and Community for Information Access Research) tasks.[17]

nDCG   Microsoft version [4] of the *normalised Discounted Cumulative Gain* [13]. Handles graded relevance. This version of nDCG has been widely used in the IR community since around 2005 [4].

nERR   A normalised version of *expected reciprocal rank* [8]. Handles graded relevance; unlike the other three measures, more suitable for navigational intents than for informational intents. Also widely used in the IR community, since its proposal in 2009.

---

[16]The *covariance* of two random variables x and y is defined as $COV(x, y) = E((x - E(x))(y - (y)))$; note that $COV(x, x) = V(x)$, i.e. the population variance of x (see Chap. 1 Sect. 1.2.1). Now, in general, $V(x - y) = V(x) + V(y) - 2COV(x, y)$ holds. However, if $COV(x, y) = 0$, we say that x and y are *uncorrelated*.

[17]http://research.nii.ac.jp/ntcir/index-en.html

Table 6.5 Some variance estimates ($\hat{\sigma}^2$) from Sakai [22]

Data	md	m_C	n_C	AP	Q	nDCG	nERR
TREC 2003 robust	1000	78	50	0.048	0.047	0.046	0.114
TREC 2004 robust	1000	78	49	0.046	0.046	0.046	0.115
Pooled	–	–	–	0.047	0.047	0.046	0.115
TREC 2003 robust	10	78	50	0.089	0.066	0.073	0.121
TREC 2004 robust	10	78	49	0.077	0.063	0.073	0.121
Pooled	–	–	–	0.084	0.065	0.073	0.121
TREC 2011 web ad hoc	10	25	50	0.088	0.048	0.054	0.101
TREC 2012 web ad hoc	10	20	50	0.077	0.026	0.034	0.072
Pooled	–	–	–	0.082	0.037	0.044	0.086

The exact definitions of all of these measures can be found in Sakai [21].

A few observations can be made from Table 6.5:

- The relative trends of the evaluation measures in terms of stability are similar across the different data sets. Specifically, nERR is consistently less stable than Q and nDCG.[18] Hence nERR requires more topics than other measures under the same statistical requirement.
- Under the same experimental setting, the variance estimates for a particular evaluation measure based on different data sets can be quite similar, especially if the topic-by-run matrices are large. For example, note that the variances obtained from the TREC 2003 and 2004 robust data with $md = 1000$ are very similar to each other.
- The variance of the same evaluation measure for the same data set is inflated as the measurement depth is reduced. For example, while the variance of AP for the TREC 2003 robust data with $md = 1000$ is 0.048, that with $md = 10$ is 0.089. This is because the latter relies on considerably fewer data points (i.e. retrieved relevant documents).

The variance of an evaluation measure also depends on the degree of *incompleteness* of relevance assessments, i.e. how many unjudged documents there are in the target corpus. See Sakai [24] for more discussions.

6.8 Comparing the Different Topic Set Size Design Methods

This section clarifies the relationships among the aforementioned three topic set size design tools, under the premise that we use $\hat{\sigma}_d^2 = 2\hat{\sigma}^2$ (i.e. the assumption that the

---

[18]The high variances of nERR reflect the fact that it is a measure designed primarily for navigational intents. That is, this measure relies heavily on the first retrieved relevant document, while the other measures rely on the other retrieved relevant documents as well.

variance of the score *differences* is twice the variance of the scores) for the paired-data tools `samplesizeTTEST2` and `samplesizeCI2`.

In Sect. 6.8.1, I first point out that one-way ANOVA with $m = 2$ systems is strictly equivalent to the two-sample t-test and compare the power-based topic set size design results based on these two approaches. Note that the minimum detectable *range* (*minD*) for the ANOVA-based tool is defined based on the best and the worst performing systems and therefore reduces to the minimum detectable *difference* (*minD_t*) for the t-test-based tools when $m = 2$. In addition, I compare the results based on paired and two-sample t-tests.

In Sect. 6.8.2, I compare the CI-based topic set size design results based on paired and unpaired data.

Based on the results discussed in Sects. 6.8.1 and 6.8.2, I recommend `samplesize2SAMPLET` and `samplesize2SAMPLECI` (i.e. tools based on unpaired data) over `samplesizeTTEST2` and `samplesizeCI2` (i.e., tools based on paired data) for the following two reasons:

(a) The unpaired versions give slightly tighter estimates while avoiding the use of $\hat{\sigma}_d^2$ (i.e. they only require $\hat{\sigma}^2$);
(b) Topic set sizes that are large enough for unpaired experiments always suffice for paired experiments as well.

Moreover, even for researchers interested in t-test-based topic set size design, I recommend `samplesizeANOVA2` with $m = 2$ over `samplesize2SAMPLET` and `samplesizeTTEST2`, as it returns the tightest topic set size design estimates despite the fact that one-way ANOVA with $m = 2$ systems is strictly equivalent to the two-sample t-test.

In Sect. 6.8.3, I compare one-way ANOVA-based topic set size results with those based on a cap on the CI width and explain why the results are almost identical under certain settings. Based on the above observation, Sakai [23] recommended the use of the ANOVA-based approach even if the researcher is interested in the CI-based approach. However, that was only because the early Excel tool `samplesizeCI` (for paired data) could not handle large topic set sizes due to the computation of c^* (Eq. 6.43 in Sect. 6.5.2). As the new CI-based topic set design tools `samplesizeCI2` (for paired data) and `samplesize2SAMPLECI` (for unpaired data) are free from this problem, researchers can just use these tools if they are interested in the expected CI width; for the reasons mentioned above, `samplesize2SAMPLECI` would be the recommended tool.

Below, we shall provide detailed discussions on the above points.

6.8.1 Paired and Two-Sample t-Tests vs. One-Way ANOVA

6.8.1.1 Two-Sample *t*-Test vs. One-Way ANOVA

Recall Corollary 7 from Chap. 1 Sect. 1.2.5: if t is a random variable that obeys $t(\phi)$, then t^2 obeys $F(1, \phi)$. Hence, if $t_{inv}(\phi; \alpha)$ is the critical value for a two-sided two-sample t-test and $F_{inv}(1, \phi, \alpha)$ is the critical value for a one-way ANOVA, $t_{inv}(\phi; \alpha)^2 = F_{inv}(1, \phi; \alpha)$ holds. On the other hand, recall the test statistics of these two tests (Chap. 2 Sect. 2.3 and Chap. 3 Sect. 3.1.2):

$$t_0 = \frac{\bar{x}_{1\bullet} - \bar{x}_{2\bullet}}{\sqrt{V_p(1/n_1 + 1/n_2)}} = \frac{\sqrt{\frac{n_1 n_2}{n_1 + n_2}}(\bar{x}_{1\bullet} - \bar{x}_{2\bullet})}{\sqrt{V_p}}, \tag{6.56}$$

where

$$V_p = \frac{S_1 + S_2}{n_1 + n_2 - 2} = \frac{\sum_{j=1}^{n_1}(x_{1j} - \bar{x}_{1\bullet})^2 + \sum_{j=1}^{n_2}(x_{2j} - \bar{x}_{2\bullet})^2}{n_1 + n_2 - 2}, \tag{6.57}$$

and

$$F_0 = \frac{V_A}{V_{E1}} = \frac{\left(\sum_{i=1}^{m} n_i (\bar{x}_{i\bullet} - \bar{x})^2\right)/\phi_A}{\left(\sum_{i=1}^{m} \sum_{j=1}^{n_i}(x_{ij} - \bar{x}_{i\bullet})^2\right)/\phi_{E1}}, \tag{6.58}$$

where $\phi_A = m - 1$, $\phi_{E1} = \sum_{i=1}^{m} = n_i - m$. From Eq. 6.56, we have:

$$t_0^2 = \frac{\frac{n_1 n_2}{n_1 + n_2}(\bar{x}_{1\bullet} - \bar{x}_{2\bullet})^2}{V_p}. \tag{6.59}$$

When $m = 2$, note that $\phi_A = 1$, $\phi_{E1} = n_1 + n_2 - 2$ and that Eq. 6.58 is reduced to the following:

$$F_0 = \frac{n_1(\bar{x}_{1\bullet} - \bar{x})^2 + n_2(\bar{x}_{2\bullet} - \bar{x})^2}{\left(\sum_{j=1}^{n_1}(x_{1j} - \bar{x}_{1\bullet})^2 + \sum_{j=1}^{n_2}(x_{2j} - \bar{x}_{2\bullet})^2\right)/(n_1 + n_2 - 2)}$$

$$= \frac{n_1\bar{x}_{1\bullet}^2 + n_1\bar{x}^2 - 2n_1\bar{x}_{1\bullet}\bar{x} + n_2\bar{x}_{2\bullet}^2 + n_2\bar{x}^2 - 2n_2\bar{x}_{2\bullet}\bar{x}}{V_p}. \tag{6.60}$$

Let us show that $t_0^2 = F_0$ when $m = 2$. To do this, from Eqs. 6.59 and 6.60, it suffices to prove the following:

$$\frac{n_1 n_2}{n_1 + n_2}(\bar{x}_{1\bullet} - \bar{x}_{2\bullet})^2 = n_1\bar{x}_{1\bullet}^2 + n_1\bar{x}^2 - 2n_1\bar{x}_{1\bullet}\bar{x} + n_2\bar{x}_{2\bullet}^2 + n_2\bar{x}^2 - 2n_2\bar{x}_{2\bullet}\bar{x}. \tag{6.61}$$

The proof is straightforward if we let $N = n_1 + n_2$ and utilise the fact that the sample grand mean for the case of $m = 2$ can be expressed as[19]:

$$\bar{x} = \frac{\sum_{j=1}^{n_1} x_{1j} + \sum_{j=1}^{n_2} x_{2j}}{N} = \frac{n_1 \bar{x}_{1\bullet} + n_2 \bar{x}_{2\bullet}}{N} . \qquad (6.62)$$

Thus, when $m = 2$, both $t_{inv}(\phi; \alpha)^2 = F_{inv}(1, \phi; \alpha)$ and $t_0^2 = F_0$ hold. In other words, these tests are the strictly equivalent: the only difference is whether both the test statistic and the critical value are squared or not.

Despite the above fact, $\mathtt{samplesizeANOVA2}$ (with $m = 2$) and $\mathtt{samplesize2SAMPLET}$ return slightly different sample sizes given the same requirements, as I shall demonstrate below. This is because we follow different paths in t-test-based and ANOVA-based topic set size designs: the former involves approximating a noncentral t-distribution, while the latter involves approximating a noncentral F-distribution. As $\mathtt{samplesizeANOVA2}$ (with $m = 2$) returns tighter topic set size estimates than $\mathtt{samplesize2SAMPLET}$, I recommend the former if the researcher prefers tight estimates.

6.8.1.2   Paired t-Test vs. Two-Sample t-Test

Let us also compare $\mathtt{samplesizeTTEST2}$ for paired data and $\mathtt{samplesize2}$ $\mathtt{SAMPLET}$ for unpaired data under the same requirements and the premise that we let $\hat{\sigma}_d^2 = 2\hat{\sigma}^2$.

From Eqs. 6.19 and 6.22, recall that the initial topic set size estimate for the paired t-test is given by:

$$n \approx \left(\frac{z_{inv}(\alpha/2) - z_{inv}(1 - \beta)}{minD_t} \right)^2 \hat{\sigma}_d^2 + \frac{z_{inv}(\alpha/2)^2}{2} . \qquad (6.63)$$

Substituting $\hat{\sigma}_d^2 = 2\hat{\sigma}^2$ to Eq. 6.63 gives us:

---

[19] Start from the left hand side of Eq. 6.61.

$$n_1 \bar{x}_{1\bullet}^2 + n_2 \bar{x}_{2\bullet}^2 - 2\bar{x}(n_1 \bar{x}_{1\bullet} + n_2 \bar{x}_{2\bullet}) + (n_1 + n_2)\bar{x}^2 = n_1 \bar{x}_{1\bullet}^2 + n_2 \bar{x}_{2\bullet}^2 - 2N\bar{x}^2 + N\bar{x}^2$$

$$= n_1 \bar{x}_{1\bullet}^2 + n_2 \bar{x}_{2\bullet}^2 - N\frac{(n_1 \bar{x}_{1\bullet} + n_2 \bar{x}_{2\bullet})^2}{N^2} = n_1 \bar{x}_{1\bullet}^2 + n_2 \bar{x}_{2\bullet}^2 - \frac{n_1^2 \bar{x}_{1\bullet}^2 + n_2^2 \bar{x}_{2\bullet}^2 + 2n_1 n_2 \bar{x}_{1\bullet} \bar{x}_{2\bullet}}{N}$$

$$= \frac{1}{N}((n_1 + n_2)n_1 \bar{x}_{1\bullet}^2 + (n_1 + n_2)n_2 \bar{x}_{2\bullet}^2 - n_1^2 \bar{x}_{1\bullet}^2 - n_2^2 \bar{x}_{2\bullet}^2 - 2n_1 n_2 \bar{x}_{1\bullet} \bar{x}_{2\bullet})$$

$$= \frac{1}{N}(n_1 n_2 \bar{x}_{1\bullet}^2 + n_1 n_2 \bar{x}_{2\bullet}^2 - 2n_1 n_2 \bar{x}_{1\bullet} \bar{x}_{2\bullet}) = \frac{n_1 n_2}{N}(\bar{x}_{1\bullet} - \bar{x}_{2\bullet})^2 ,$$

which equals the right hand side of Eq. 6.61.

$$ n \approx 2\,(z_{inv}(\alpha/2) - z_{inv}(1-\beta))^2 \,\frac{\hat{\sigma}^2}{minD_t^2} + \frac{z_{inv}(\alpha/2)^2}{2}\,. \tag{6.64} $$

On the other hand, from Eqs. 6.27 and 6.28, the initial topic size estimate for the two-sample t-test is given by:

$$ n \approx 2\,(z_{inv}(\alpha/2) - z_{inv}(1-\beta))^2 \,\frac{\hat{\sigma}^2}{minD_t^2} + \frac{z_{inv}(\alpha/2)^2}{4}\,. \tag{6.65} $$

By comparing Eqs. 6.64 and 6.65, the initial estimate for the paired t-test is actually larger than that for the two-sample t-test by exactly $z_{inv}(\alpha/2)^2/4$, which is about 0.96 (i.e. the topic set sizes differ by about only one topic) when $\alpha = 0.05$. Thus, even though paired t-test should require fewer topics than the unpaired t-test, our use of $\hat{\sigma}_d^2 = 2\hat{\sigma}^2$ makes the two topic set size design results almost identical, as we shall demonstrate below with real data.

6.8.1.3   Power-Based Topic Set Size Design Results

Table 6.6 compares the topic set size design results based on the paired t-test, two-sample t-test, and one-way ANOVA ($m = 2$), using the pooled variance estimates shown in Table 6.5, with $(\alpha, \beta) = (0.05, 0.20)$. Recall that the minimum detectable *range* (*minD*) for one-way ANOVA reduces to the minimum detectable *difference* (*minD$_t$*) for t-tests when $m = 2$. For example, since the pooled variance for AP from the TREC robust track data with measurement depth 1000 is $\hat{\sigma}^2 = 0.047$, we let $\hat{\sigma}_d^2 = 2 * 0.047 = 0.094$. For *minD$_t$* $= 0.05$, `samplesizeTTEST2` returns 298 topics; `samplesize2SAMPLET` returns 297 topics; and `samplesizeANOVA2` returns 289 topics. As indicated in bold, it can be observed that `samplesizeANOVA2` with $m = 2$ gives us the tighest estimates and that `samplesizeTTEST2` gives us the least tight estimates due to our use of $\hat{\sigma}_d^2 = 2\hat{\sigma}^2$. Thus, I recommend the use of `samplesizeANOVA2` over `samplesizeTTEST2` and `samplesize2SAMPLET`; and if the researcher needs to choose between the two t-test based tools, I recommend `samplesize2SAMPLET`. Sample sizes that are appropriate for unpaired data are always large enough for paired data.

Below, I compare the *initial* topic set size estimates based on the t-tests and one-way ANOVA and show that the tight estimates obtained from ANOVA are not necessarily due to the *initial* estimates but rather due to the checking of the achieved power. Recall that Table 6.6 shows the results of automatically testing *several* topic set sizes against the power requirement and selecting the minimal one. As was mentioned in Sect. 6.4.1, the initial topic set size estimate of `samplesizeANOVA2` is rather rough, and the final topic set size can be substantially *smaller*, giving us a tight estimate.

Table 6.6 Power-based topic set size design results: paired t-test/two-sample t-test/one-way ANOVA ($m = 2$), based on the variance estimates shown in Table 6.5, with $(\alpha, \beta) = (0.05, 0.20)$. In each cell, the tightest estimate is shown in bold

$minD(_t)$	AP	Q	nDCG	nERR
TREC 2003+2004 robust ($md = 1000$)				
0.05	298/297/**289**	298/297/**289**	291/290/**283**	725/724/**705**
0.10	76/75/**73**	76/75/**73**	75/74/**71**	183/182/**177**
0.15	35/34/**33**	35/34/**33**	35/34/**32**	83/82/**79**
0.20	21/20/**19**	21/20/**19**	21/20/**19**	48/47/**45**
0.25	14/13/**12**	14/13/**12**	14/13/**12**	31/30/**29**
TREC 2011+2012 web ad hoc ($md = 10$)				
0.05	517/516/**503**	235/234/**227**	279/278/**270**	542/541/**527**
0.10	131/130/**126**	61/60/**58**	72/71/**68**	137/136/**133**
0.15	60/59/**57**	28/27/**26**	33/32/**31**	62/61/**59**
0.20	35/34/**32**	17/16/**15**	20/19/**18**	36/35/**34**
0.25	23/22/**21**	12/11/**10**	14/13/**12**	24/23/**22**

Table 6.7 Linear approximations of λ, the noncentrality parameter of a noncentral χ^2 distribution $\chi'^2(\phi_A, \lambda)$, when $\phi_A = 2$

β	$\alpha = 0.01$	$\alpha = 0.05$
0.1	15.652	11.293
0.2	12.287	8.444
0.3	10.107	6.634
0.4	8.408	5.255
0.5	6.955	4.106

The initial topic set size estimates for the paired and two-sample t-tests in the present considerations are given by Eqs. 6.64 and 6.65. As for one-way ANOVA, the initial estimate is given by (Sect. 6.4.2):

$$n \approx 2\lambda \frac{\hat{\sigma}^2}{minD^2} , \tag{6.66}$$

where $minD$ is the same as $minD_t$ when $m = 2$ as was explained earlier. As for λ, since $\phi_A = 1$ when $m = 2$, Table 6.4 is reduced to Table 6.7; since we are considering $(\alpha, \beta) = (0.05, 0.20)$ here, $\lambda = 11.293$.

Table 6.8 shows the initial topic set sizes that correspond to the final results shown in Table 6.6; these integers were obtained by rounding up the numbers in the initial estimate cells of `samplesizeTTEST2`, `samplesize2SAMPLET`, and `samplesizeANOVA2` ($m = 2$) with $(\alpha, \beta) = (0.05, 0.20)$. The initial estimates in the tools are computed as in Eqs. 6.64, 6.65, and 6.66 with $\lambda = 11.293$. It can be observed that while the initial estimates for the t-tests are almost identical to the final topic set sizes shown in Table 6.6, the initial estimates for one-way ANOVA are larger than the final topic set sizes especially for small $minD$s. As a result, in Table 6.8, the initial estimates with `samplesize2SAMPLET` are generally smaller than those with `samplesizeANOVA2`. This demonstrates the importance

Table 6.8 The initial topic set sizes estimates for paired t-test/two-sample t-test/one-way ANOVA ($m = 2$), based on the variance estimates shown in Table 6.5, with $(\alpha, \beta) = (0.05, 0.20)$. In each cell, the tightest estimate is shown in bold. This table should be compared with the *final* topic set size results given in Table 6.6

$minD_{(t)}$	AP	Q	nDCG	nERR
TREC 2003+2004 robust ($md = 1000$)				
0.05	**297/297**/318	**297/297**/318	291/**290**/311	**724/724**/777
0.10	76/**75**/80	76/**75**/80	75/**74**/78	183/**182**/195
0.15	35/**34**/36	35/**34**/36	**34/34**/35	83/**82**/87
0.20	21/**20/20**	21/**20/20**	20/**19**/20	48/**47**/49
0.25	14/**13/13**	14/**13/13**	14/**13/13**	31/**30**/32
TREC 2011+2012 web ad hoc ($md = 10$)				
0.05	517/**516**/554	235/**234**/250	279/**278**/298	542/**541**/581
0.10	131/**130**/139	60/**59**/63	71/**70**/75	137/**136**/146
0.15	**60/60**/62	28/**27**/28	33/**32**/33	62/**61**/65
0.20	35/**34**/35	17/**16/16**	20/**19/19**	36/**35**/37
0.25	23/**22**/23	12/11/**10**	13/**12/12**	24/**23**/24

of actually checking the achieved power: `samplesizeANOVA2` starts with a rough estimate given by Eq. 6.66 but ends up returning a tighter topic set size by utilising Eq. 6.34.

6.8.2   CI-Based Topic Set Size Design: Paired vs. Unpaired Data

Let us now compare the CI-based topic set size design results with paired-data and unpaired-data settings using the pooled variances shown in Table 6.5.

Under $\hat{\sigma}_d^2 = 2\hat{\sigma}^2$, the initial topic set size estimate for the paired data case given by Eq. 6.47 can be rewritten as:

$$n \approx \frac{8z_{inv}(\alpha/2)^2 \hat{\sigma}^2}{\delta^2} , \tag{6.67}$$

which is actually identical to the inequality for the unpaired data case given by Eq. 6.51. That is, the initial estimates in `samplesizeCI2` and `samplesize2SAMPLECI` are actually the same.

What slightly differ are the final constraints for checking the expected CI width, as discussed below. Under $\hat{\sigma}_d^2 = 2\hat{\sigma}^2$, the constraint for the paired data case given by Eq. 6.45 can be rewritten as:

$$\frac{t_{inv}(\phi; \alpha)(1 - 1/4\phi)}{\sqrt{n}} \leq \frac{\delta}{2\sqrt{2}\sigma} , \quad \phi = n - 1 . \tag{6.68}$$

Table 6.9 95% CI-based topic set size design results using the pooled variances from Table 6.5: paired/unpaired data. In each cell, the tightest estimate is shown in bold

δ	AP	Q	nDCG	nERR
\multicolumn{5}{c}{TREC 2003+2004 robust ($md = 1000$)}				
0.10	147/**146**	147/**146**	144/**143**	356/**355**
0.20	39/**38**	39/**38**	38/**37**	91/**90**
0.30	19/**18**	19/**18**	18/**17**	42/**41**
0.40	12/**11**	12/**11**	11/**10**	25/**24**
0.50	8/**7**	8/**7**	8/**7**	17/**16**
\multicolumn{5}{c}{TREC 2011+2012 web ad hoc ($md = 10$)}				
0.10	254/**253**	116/**115**	138/**137**	267/**266**
0.20	65/**64**	31/**30**	36/**35**	69/**68**
0.30	30/**29**	15/**14**	18/**17**	32/**31**
0.40	18/**17**	10/**9**	11/**10**	19/**18**
0.50	13/**12**	7/**6**	8/**7**	13/**12**

This is the same as the constraint for the unpaired data case given by Eq. 6.49, except that the unpaired data case uses $\phi = n_1 + n_2 - 2 = 2n - 2$. Hence the topic set sizes returned by `samplesizeCI2` and `samplesize2SAMPLECI` are naturally almost identical.

Table 6.9 shows the 95%CI-based topic set size design results: it can be observed that the two tools indeed return almost identical numbers and that `samplesize2SAMPLECI` returns slightly higher estimates, as indicated in bold. For the reasons discussed earlier, I recommend `samplesize2SAMPLECI` over `samplesizeCI2`.

6.8.3 One-Way ANOVA vs. Confidence Intervals

This section explains why `samplesizeANOVA2` and the CI-based tools `samplesizeCI2` and `samplesize2SAMPLECI` return almost identical topic set sizes under certain conditions [23]. Recall that `samplesizeANOVA2` requires $(\alpha, \beta, m, minD, \hat{\sigma}^2)$ and outputs the topic set size that achieves the desired statistical power (Sect. 6.2.1), whereas `samplesizeCI2` requires $(\alpha, \delta, \hat{\sigma}_d^2)$ and `samplesize2SAMPLECI` requires $(\alpha, \delta, \hat{\sigma}^2)$, to output the topic set size for a given cap on the expected CI width of the difference between any two systems (Sect. 6.5.1).

As we have discussed in Sect. 6.8.2, the *initial* topic set size for both of the CI-based tools is given by Eq. 6.67; let n_{CI} denote this initial estimate. On the other hand, the initial estimate for one-way ANOVA is given by Eq. 6.66; let n_{ANOVA} denote this. Then we have:

$$\frac{n_{ANOVA}}{n_{CI}} = \frac{\lambda}{4z_{inv}(\alpha/2)^2}\left(\frac{\delta}{minD}\right)^2. \tag{6.69}$$

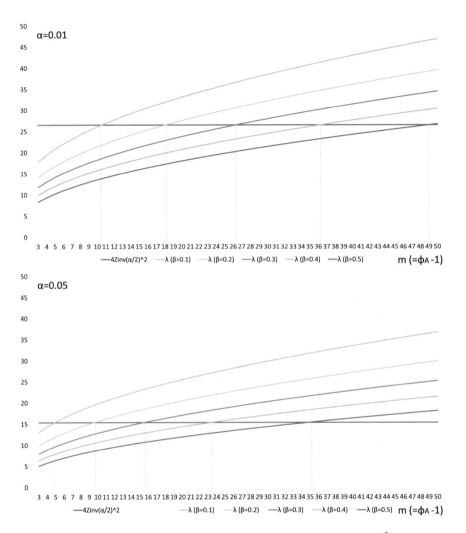

Fig. 6.6 The relationship between the λ of `samplesizeANOVA2` and $4z_{inv}(\alpha/2)^2$

Note that $4z_{inv}(\alpha/2)^2$ is a constant for a given α and that λ is a constant given (α, β, m): see Table 6.4.

Figure 6.6 visualises the relationship between $4z_{inv}(\alpha/2)^2$ and λ while varying the number of systems m for λ. It can be observed that

$$\lambda \approx 4z_{inv}(\alpha/2)^2 \tag{6.70}$$

holds when one of the following conditions holds:

Condition A $(\alpha, \beta, m) = (0.01, 0.10, 10)$;

Condition B $(\alpha, \beta, m) = (0.01, 0.20, 18)$;
Condition C $(\alpha, \beta, m) = (0.01, 0.30, 26)$;
Condition D $(\alpha, \beta, m) = (0.01, 0.40, 36)$;
Condition E $(\alpha, \beta, m) = (0.01, 0.50, 49)$;
Condition F $(\alpha, \beta, m) = (0.05, 0.10, 5)$;
Condition G $(\alpha, \beta, m) = (0.05, 0.20, 10)$;
Condition H $(\alpha, \beta, m) = (0.05, 0.30, 16)$;
Condition I $(\alpha, \beta, m) = (0.05, 0.40, 24)$;
Condition J $(\alpha, \beta, m) = (0.05, 0.50, 35)$.

Hence, whenever one of Conditions A-J holds, from Eqs. 6.69 and 6.70, we obtain:

$$\frac{n_{ANOVA}}{n_{CI}} \approx \left(\frac{\delta}{minD} \right)^2. \tag{6.71}$$

Therefore, if we let $\delta = minD$ under one of the above ten conditions, we obtain $n_{ANOVA}/n_{CI} \approx 1$, or $n_{ANOVA} \approx n_{CI}$, regardless of the variance estimate $\hat{\sigma}^2$. However, recall that these are the *initial* topic set size estimates, not the final ones returned by these tools.

Based on the above observation, Sakai [23] recommended using the ANOVA-based approach even if the researcher is interested in a CI-based constraint, as an early version of the CI-based tool `samplesizeCI` could not handle large topic set sizes due to its reliance on the gamma function. However, the new tools `samplesizeCI2` and `samplesize2SAMPLECI` are free from this problem. As was discussed earlier, if the researcher is interested in CI-based topic set size design, I recommend `samplesize2SAMPLECI`: use the "`approximation`" sheet, not the "`using gamma`" sheet: the approximation of c^* is highly accurate, as was discussed in Sect. 6.5.2.

References

1. J. Allan, B. Carterette, J.A. Aslam, V. Pavlu, B. Dachev, E. Kanoulas, Million query track 2007 overview, in *Proceedings of TREC 2007*, Gaithersburg, 2008
2. J. Allan, J.A. Aslam, B. Carterette, V. Pavlu, E. Kanoulas, Million query track 2008 overview, in *Proceedings of TREC 2008*, Gaithersburg, 2009
3. C. Buckley, E.M. Voorhees, Retrieval system evaluation, in *TREC: Experiment and Evaluation in Information Retrieval*, ed. by E.M. Voorhees, D.K. Harman, chapter 3, pp. 53–75 (The MIT Press, Cambridge, MA, 2005)
4. C. Burges, T. Shaked, E. Renshaw, A. Lazier, M. Deeds, N. Hamilton, G. Hullender, Learning to rank using gradient descent, in *Proceedings of ACM ICML*, Bonn, 2005, pp. 89–96
5. B. Carterette, J. Allan, R. Sitaraman, Minimal test collections for retrieval evaluation, in *Proceedings of ACM SIGIR*, Seattles, 2006, pp. 268–275
6. B. Carterette, V. Pavlu, E. Kanoulas, J.A. Aslam, J. Allan, Evaluation over thousands of queries, in *Proceedings of ACM SIGIR*, Singapore, 2008, pp. 651–658
7. B. Carterette, V. Pavlu, H. Fang, E. Kanoulas, Million query track 2009 overview, in *Proceedings of TREC 2009*, Gaithersburg, 2010

8. O. Chapelle, D. Metzler, Y. Zhang, P. Grinspan, Expected reciprocal rank for graded relevance, in *Proceedings of ACM CIKM*, Hong Kong, 2009, pp. 621–630
9. C.L.A. Clarke, N. Craswell, I. Soboroff, E.M. Voorhees, Overview of the TREC 2011 web track, in *Proceedings of TREC 2011*, Gaithersburg, 2012
10. C.L.A. Clarke, N. Craswell, E.M. Voorhees, Overview of the TREC 2012 web track, in *Proceedings of TREC 2012*, Gaithersburg, 2013
11. H. Gilbert, K. Sparck Jones, Statistical bases of relevance assessment for the 'ideal' information retrieval test collection. Technical report, Computer Laboratory, University of Cambridge, British Library Research and Development Report No. 5481 (1979)
12. D.K. Harman, The TREC test collections, in *TREC: Experiment and Evaluation in Information Retrieval*, ed. by E.M. Voorhees, D.K. Harman, chapter 2 (The MIT Press, Cambridge, MA, 2005)
13. K. Järvelin, J. Kekäläinen, Cumulated gain-based evaluation of IR techniques. ACM TOIS **20**(4), 422–446 (2002)
14. H.C. Kraemer, C. Blasey, *How Many Subjects? Statistical Power Analysis in Research*, 2nd edn. (SAGE Publications, Los Angeles, 2016)
15. K.R. Murphy, B. Myors, A. Wolach, *Statistical Power Analysis: A Simple and General Model for Traditional and Modern Hypothesis Tests*, 4th edn. (Routledge, London, 2014)
16. Y. Nagata, *How to Design the Sample Size (in Japanese)* (Asakura Shoten, Shinjuku, 2003)
17. Y. Nagata, M. Yoshida, *Introduction to Multiple Comparison Procedures (in Japanese)* (Scientist Press, Shibuya, 1997)
18. T.P. Ryan, *Sample Size Determination and Power* (Wiley, Chichester, 2013)
19. T. Sakai, Ranking the NTCIR systems based on multigrade relevance, in *Proceedings of AIRS 2004*, Beijing. LNCS 3411, 2004, pp. 251–262
20. T. Sakai, Evaluating evaluation metrics based on the bootstrap, in *Proceedings of ACM SIGIR*, Seattle, 2006, pp. 525–532
21. T. Sakai, Metrics, statistics, tests, in *PROMISE Winter School 2013: Bridging between Information Retrieval and Databases (LNCS 8173)*, 2014, pp. 116–163
22. T. Sakai, Topic set size design. Inf. Retr. **19**(3), 256–283 (2016)
23. T. Sakai, Evaluating evaluation measures with worst-case confidence interval widths, *in Proceedings of EVIA*, Chiyoda, 2017, pp. 16–19
24. T. Sakai, How to run an evaluation task, in *Information Retrieval Evaluation in a Changing World: Lessons Learned from 20 Years of CLEF*, ed. by N. Ferro, C. Peters, chapter 3. (Springer, 2019)
25. T. Sakai, L. Shang, On estimating variances for topic set size design, in *Proceedings of EVIA*, Chiyoda, 2016, pp. 9–12
26. M. Sanderson, J. Zobel, Information retrieval evaluation: effort, sensitivity, and reliability, in *Proceedings of ACM SIGIR*, Salvador, 2005, pp. 162–169
27. K. Sparck Jones, C.J. van Rijsbergen, Report on the need for and provision of an 'ideal' information retrieval test collection. Technical report, Computer Laboratory, University of Cambridge, British Library Research and Development Report No. 5266, 1975
28. K. Sparck Jones, R.G. Bates, Report on a design study for the 'ideal' information retrieval test collection. Technical report, Computer Laboratory, University of Cambridge, British Library Research and Development Report No. 5481, 1977
29. E.M. Voorhees, Overview of the TREC 2003 robust retrieval track, in *Proceedings of TREC 2003*, Gaithersburg, 2004
30. E.M. Voorhees, Overview of the TREC 2004 robust retrieval track, in *Proceedings of TREC 2004*, Gaithersburg, 2005
31. E.M. Voorhees, Topic set size redux, in *Proceedings of ACM SIGIR*, Boston, 2009, pp. 806–807
32. E.M. Voorhees, C. Buckley, The effect of topic set sizes on retrieval experiment error, in *Proceedings of ACM SIGIR*, Tampere, 2002, pp. 162–169
33. W. Webber, A. Moffat, J. Zobel, Statistical power in retrieval experimentation, in *Proceedings of ACM CIKM*, Napa Valley, 2008, pp. 571–580
34. J. Zobel, How reliable are the results of large-scale information retrieval experiments? in *Proceedings of ACM SIGIR*, Melbourne, 1998, pp. 307–314

Chapter 7
Power Analysis Using R

Abstract This section describes how power analysis on published papers can be done using a suite of simple R scripts, so that better-designed experiments can be conducted in the future. Here, "better" means "ensuring appropriate statistical power". First, an overview of the five R scripts is given (Sect. 7.2), followed by a description of each script (Sects. 7.3, 7.4, 7.5, 7.6, and 7.7). The five scripts, which are for paired t-test, two-sample t-test, one-way ANOVA, two-way ANOVA without replication, and two-way ANOVA with replication, respectively, were adapted from the R scripts of Toyoda (Introduction to statistical power analysis: a tutorial with R (in Japanese). Tokyo Tosyo, 2009): his original scripts, which contain Japanese character codes, are available from his book's website (http://www.tokyo-tosho. co.jp/download/DL02065.zip); Toyoda's scripts (and therefore mine as well) rely on R libraries called `stats` and `pwr`. (The present author is solely responsible for any problems caused by modifying the original scripts of Toyoda.) Finally, it provides summary while touching upon a survey I conducted using these R scripts, with a decade's worth of IR papers from ACM SIGIR (http://sigir.org/) and TOIS (https://tois.acm.org/) (Sakai Statistical significance, power, and sample sizes: a systematic review of SIGIR and TOIS. In: Proceedings of ACM SIGIR 2016, pp 5–14, 2016), where it was demonstrated that there are highly overpowered and highly underpowered experiments in the results reported in the IR literature. Highly overpowered experiments use a lot more resources than necessary, while highly underpowered experiments are highly likely to miss important differences that exist due to the use of small samples. We can probably do better by learning from previous studies and/or from pilot studies.

Keywords One-way ANOVA · Paired t-test · Power analysis · Student's t-test · Two-way ANOVA

7.1 Introduction

Having already looked closely at how statistical power can be computed in the context of topic set size design in Chap. 6 (for t-tests and one-way ANOVA only),

© Springer Nature Singapore Pte Ltd. 2018

T. Sakai, *Laboratory Experiments in Information Retrieval*,

The Information Retrieval Series 40, https://doi.org/10.1007/978-981-13-1199-4_7

this chapter discusses some R scripts for power analysis and the `stats/pwr` libraries of R as black boxes. That is, we only discuss how they can be utilised and how the results should be interpreted. Details of the functions provided in `stats/pwr` can be found in R's online manual.

7.2 Overview of the R Scripts for Power Analysis

The R scripts discussed in the remainder of this chapter can be downloaded from https://waseda.box.com/SIGIR2016PACK: unzip the file, and use the R code `power_code_sakai.R`. These scripts are for examining the *achieved power* and the *effect sizes* of a reported parametric statistical significance test (namely, a *t*-test or an ANOVA test) and for estimating appropriate sample sizes for a similar future experiment for achieving a *desired power*.

The above R code contains the following R functions [3]:

`future.sample.pairedt`   Given a paired *t*-test result, this script takes as input the *t*-statistic, the actual sample size, and the desired combination of (α, β) and computes the achieved power and an effect size, as well as the required sample size for achieving the desired power (Sect. 7.3).

`future.sample.unpairedt`   This is similar to the above, except that it is for a two-sample (student's) *t*-test and therefore takes two-sample sizes with a *t*-statistic as input (Sect. 7.4).

`future.sample.1wayanova`   Given a one-way ANOVA result, this script takes as input the *F*-statistic, the number of systems *m*, and the number of topics *n* and computes the achieved power and an effect size, as well as the required sample size for achieving the desired power (Sect. 7.5).

`future.sample.2waynorep`   This is similar to the above, except that it is for two-way ANOVA without replication (Sect. 7.6).

`future.sample.2wayanova2`   This is similar to the above, except that it is for two-way ANOVA with replication. It computes the achieved power and the required sample size for the system effect, for the topic effect, and also for the interaction between the two (Sect. 7.7).

Below, each script is discussed in turn. It should be noted that all the examples provided below were taken from a survey I conducted on ACM SIGIR and TOIS papers [3] and are *real* examples from IR literature.

7.3 Power Analysis with the Paired *t*-Test

Figure 7.1 shows the source code of `future.sample.pairedt`, which can be used for a given paired *t*-test result. As can be seen, this R script requires the following arguments:

```
> future.sample.pairedt
function( TSTAT, N, TAIL="two.sided", ALPHA=.05,
POW=.8 ){

cat( "INPUT: t=", TSTAT, "n=", N, "Alt=", TAIL, "ALPHA=", ALPHA,
"POWER=", POW, "\n" )
cat( "\n" )

ES <- abs( TSTAT )/sqrt( N )
cat( "EShat=", ES, "\n" )
cat("\n")

ACPOWER <- power.t.test( n=N, delta=ES, sig.level=ALPHA, type='paired',
alternative=TAIL, strict=T )$power
cat( "Achieved_power=", ACPOWER, "\n" )
cat("\n")

SAMPLESIZ <- ceiling( power.t.test(power=POW, delta=ES, sig.level=ALPHA,
type='paired', strict=T, alternative=TAIL )$n )
  cat( "Sample size n per group =", SAMPLESIZ, "\n")

}
```

Fig. 7.1 Source code of `future.sample.pairedt`

TSTAT   The *t*-statistic (t_0) reported;
N   The actual sample size *n* (e.g. number of topics);
TAIL   Whether the test is two-sided (default) or one-sided ("`one.sided`");
ALPHA α, the desired Type I error probability (default: 0.05);
POWER $1 - \beta$, the desired statistical power (default: 0.80).

From the code, it can be observed that the effect size is computed as:

$$\text{EShat} = \frac{|t_0|}{\sqrt{n}} = |d_{paired}| , \tag{7.1}$$

where d_{paired} is the standardised mean difference for paired data (Eqs. 5.2 and 5.3 from Chap. 5, Sect. 5.2.1). It can also be observed that the achieved power and the required sample size for achieving the desired power are obtained by relying on a function called `power.t.test` (from the `stats` library).[1]

Figure 7.2 shows an example where $t_0 = 0.953$, $n = 28$, with the defaults used for the other arguments. It can be observed that $|d_{paired}| = 0.1801$ and that the achieved power is very low (about 15%): this is a highly *underpowered* experiment. That is, there is an 85% probability that the researcher will miss a true difference. Thus, to study the effect of this magnitude in a future experiment under Cohen's five-eighty convention (i.e. ensuring 80% power with 5% Type I error rate), the researcher should prepare about 244 topics, instead of 28.

---

[1] Note that you can look up the specification of any standard R function using ? on the R command line, e.g. ? `'power.t.test'`.

```
> future.sample.pairedt( 0.953, 28 )
INPUT: t= 0.953 n= 28 Alt= two.sided ALPHA= 0.05 POWER= 0.8

EShat= 0.1801001

Achieved_power= 0.1510342

Sample size n per group = 244
```

Fig. 7.2 Using `future.sample.pairedt`

```
> future.sample.unpairedt
function( TSTAT, N1, N2, TAIL="two.sided",
ALPHA=.05, POW=.8 ){

cat( "INPUT: t=", TSTAT, "n1=", N1, "n2=", N2, "Alt=", TAIL, "ALPHA=",
ALPHA, "POWER=", POW, "\n" )
cat( "\n" )

ES <- abs( TSTAT )*sqrt( (N1+N2)/(N1*N2) )
cat( "EShat=", ES, "\n" )
cat("\n")

if( N1 == N2 ){
ACPOWER <- power.t.test( n=N1, delta=ES, sig.level=ALPHA,
type='two.sample', alternative=TAIL, strict=T )$power
} else {
ACPOWER <- pwr.t2n.test( n1=N1, n2=N2, d=ES, sig.level=ALPHA,
alternative=TAIL )$power
}
cat( "Achieved_power=", ACPOWER, "\n" )
cat("\n")

SAMPLESIZ <- ceiling( power.t.test(power=POW, delta=ES, sig.level=ALPHA,
type='two.sample', strict=T, alternative=TAIL )$n )
  cat( "Sample size n per group =", SAMPLESIZ, "\n")

}
```

Fig. 7.3 Source code of `future.sample.unpairedt`

7.4   Power Analysis with the Two-Sample t-Test

Figure 7.3 shows the source code of `future.sample.unpairedt`, which can be used for a given (Student's) two-sample t-test result. As can be seen, this R script requires the following arguments:

TSTAT   The t-statistic (t_0) reported;
N1, N2   The actual sample sizes n_1, n_2 (e.g. sizes of the two topic sets);
TAIL   Whether the test is two-sided (default) or one-sided ("one.sided");
ALPHA α, the desired Type I error probability (default: 0.05);
POWER $1 - \beta$, the desired statistical power (default: 0.80).

From the code, it can be observed that the effect size is computed as:

```
> future.sample.unpairedt( 2.81, 150610, 37795 )
INPUT: t= 2.81 n1= 150610 n2= 37795 Alt= two.sided ALPHA= 0.05 POWER= 0.8

EShat= 0.01616622

Achieved_power= 0.8023444

Sample size n per group = 60066
```

Fig. 7.4 Using `future.sample.unpairedt`

$$\text{EShat} = |t_0|\sqrt{\frac{n_1 + n_2}{n_1 n_2}} = |g|\,, \tag{7.2}$$

where g is Hedge's g (Eqs. 5.6 and 5.7 from Chap. 5, Sect. 5.2.1). Recall that Hedge's g is a standardised mean difference computed based on a pooled variance V_p. It can also be observed that while the power calculation relies on the aforementioned function `power.t.test` (from the `stats` library) when the two sample sizes are equal, it relies on a different function called `pwr.t2n.test` (from the `pwr` library) for unequal sample sizes. On the other hand, the required sample size for achieving the desired power is computed using `power.t.test`, under the assumption that the two future sample sizes will be equal.

Figure 7.4 shows an example where $t_0 = 2.81, n_1 = 150, 610, n_2 = 37, 795$, with the defaults used for the other arguments. It can be observed that Hedge's g is 0.0161 and that the achieved power is about 80%. Hence this is a well-designed experiment in terms of statistical power. For future experiments, the recommended sample size under the balanced design constraint ($n_1 = n_2$) would be about 60,066 per topic set.

7.5 Power Analysis with One-Way ANOVA

Figure 7.5 shows the source code of `future.sample.1wayanova`, which can be used for a given one-way ANOVA result with equal group sizes. As can be seen, this R script requires the following arguments:

FSTAT   The F-statistic (F_0) reported;
M   Number of groups (e.g. systems) m;
N   The sample size (e.g. number of topics) n per group[2];
ALPHA α, the desired Type I error probability (default, 0.05);
POWER $1 - \beta$, the desired statistical power (default, 0.80).

From the code, it can be observed that the following is computed first:

---

[2]If the group sizes are unequal, the average group size over the m groups can be used [4].

```
> future.sample.1wayanova
function( FSTAT, M, N, ALPHA=.05, POW=.8 ){

cat( "INPUT: F=", FSTAT, "M=", M, "N=", N, "ALPHA=", ALPHA, "POWER=",
POW, "\n" )
cat( "\n" )

FHAT <- sqrt( FSTAT*(M-1)/(M*(N-1)) )
cat( "Fhat=", FHAT, "\n" )
cat( "\n" )

ACPOWER <- pwr.anova.test( k=M, n=N, f=FHAT, sig.level=ALPHA )$power
cat( "Achieved_power=", ACPOWER, "\n" )
cat( "\n" )

SAMPLESIZ <- ceiling( pwr.anova.test( k=M, power=POW, f=FHAT,
sig.level=ALPHA )$n )
  cat( "Sample size n per group =", SAMPLESIZ, "\n")

}
```

Fig. 7.5 Source code of `future.sample.1wayanova`

$$\text{Fhat} = \sqrt{\frac{\phi_A F_0}{\phi_{E1}}} , \tag{7.3}$$

where $\phi_A = m - 1$, $\phi_{E1} = m(n - 1)$ and F_0 is the one-way ANOVA F-statistic as given by Chap. 3, Sect. 3.1.1, and Eqs. 3.17 and 3.18. Now, from Eqs. 3.17 and 3.18, $F_0 = \frac{\phi_{E1} S_A}{\phi_A S_{E1}}$ and therefore

$$\text{Fhat} = \sqrt{\frac{S_A}{S_{E1}}} . \tag{7.4}$$

Furthermore, the above can be rewritten using Eq. 5.20 from Chap. 5, Sect. 5.2.3 as:

$$\text{Fhat} = \sqrt{\frac{\hat{\eta}^2}{1 - \hat{\eta}^2}} . \tag{7.5}$$

The above is a form of effect size, often denoted by $\hat{f}$ [4], for this is an estimate of Cohen's *effect size index f* [1].

It can also be observed that `future.sample.1wayanova` relies on a function called `pwr.anova.test` (from the `pwr` library) to compute the achieved power and the required sample size per group for achieving the desired power.

Figure 7.6 shows an example where $F_0 = 1.284$, $m = 3$, $n = 12$, with the defaults used for the other arguments. It can be observed that $\hat{f} = 0.2790$ and that this was a highly underpowered experiment (with about 28% power). Instead of having only 12 observations (e.g. topics) per group, the researcher should have 43 per group in order to achieve 80% power.

```
> future.sample.1wayanova( 1.284, 3, 12 )
INPUT: F= 1.284 M= 3 N= 12 ALPHA= 0.05 POWER= 0.8

Fhat= 0.2789591

Achieved_power= 0.2791707

Sample size n per group = 43
```

Fig. 7.6 Using `future.sample.1wayanova`

```
> future.sample.2waynorep
function( FSTAT, M, N, ALPHA=.05, POW=.8 ){

cat( "INPUT: F=", FSTAT, "M=", M, "N=", N, "ALPHA=", ALPHA, "POWER=",
POW, "\n" )
cat( "\n" )

FHAT2 <- FSTAT/(N-1)
cat( "Fhat^2=", FHAT2, "\n" )
cat( "\n" )

ACPOWER <- pwr.f2.test( u=M-1, v=(M-1)*(N-1), FHAT2,
sig.level=ALPHA )$power
cat( "Achieved_power=", ACPOWER, "\n" )
cat( "\n" )

SAMPLESIZ <- ceiling( ( pwr.f2.test(u=M-1, power=POW, f2=FHAT2,
sig.level=ALPHA )$v + N-1 + M-1 + 1)/M )
  cat( "Sample size n per group =", SAMPLESIZ, "\n")

}
```

Fig. 7.7 Source code of `future.sample.2waynorep`

7.6   Power Analysis with Two-Way ANOVA Without Replication

Figure 7.7 shows the source code of `future.sample.2waynorep`, which can be used for a given two-way ANOVA without replication result. As can be seen, this R script requires the following arguments:

FSTAT   The F-statistic (F_0) reported;
M   Number of groups (e.g. systems) m;
N   The sample size (e.g. number of topics) n per group;
ALPHA α, the desired Type I error probability (default: 0.05);
POWER $1 - \beta$, the desired statistical power (default: 0.80).

From the code, it can be observed that the following is computed first:

$$\text{Fhat}\,\hat{}\,2 = \frac{\phi_A F_0}{\phi_{E2}} \,, \tag{7.6}$$

where $\phi_A = m - 1, \phi_{E2} = (m - 1)(n - 1)$ and F_0 is the test statistic for two-way ANOVA without replication as given by Chap. 3, Sect. 3.2, and Eqs. 3.31 and 3.32. Now, from Eqs. 3.31 and 3.32, $F_0 = \frac{\phi_{E2} S_A}{\phi_A S_{E2}}$ and therefore

$$\text{Fhat}^\wedge 2 = \frac{S_A}{S_{E2}} . \tag{7.7}$$

Furthermore, the above can be rewritten using Eq. 5.26 from Chap. 5, Sect. 5.2.3 as:

$$\text{Fhat}^\wedge 2 = \frac{\hat{\eta}_p^2}{1 - \hat{\eta}_p^2} . \tag{7.8}$$

The above is a form of *partial* effect size, denoted by $\hat{f}^2$ [4].

It can also be observed that future.sample.2waynorep relies on a function called pwr.f2.test (from the pwr library) to compute the achieved power and the required sample size per group for achieving the desired power, rather than the aforementioned pwr.anova.test.

Figure 7.8 shows an example where $F_0 = 0.63, m = 4, n = 17$, with the defaults used for the other arguments. It can be observed that $\hat{f}^2 = 0.0394$ and that this was a highly underpowered experiment (with about 18% power). Instead of having only 17 observations (e.g. topics) per group, the researcher should have 75 per group in order to achieve 80% power.

7.7 Power Analysis with Two-Way ANOVA with Replication

Figures 7.9 and 7.10 show the source code of future.sample.2wayanova2, which can be used for a given two-way ANOVA with replication result. As can be seen, this R script requires the following arguments:

FSTATA   The *F*-statistic (F_0) reported for the system effect;
FSTATB   The *F*-statistic (F_0) reported for the topic effect;
FSTATAB   The *F*-statistic (F_0) reported for the interaction (optional);
M   Number of systems m;

```
> future.sample.2waynorep( 0.63, 4, 17 )
INPUT: F= 0.63 M= 4 N= 17 ALPHA= 0.05 POWER= 0.8

Fhat^2= 0.039375

Achieved_power= 0.1834122

Sample size n per group = 75
```

Fig. 7.8 Using future.sample.2waynorep

```
> future.sample.2wayanova2
function( FSTATA, FSTATB, FSTATAB=NA, M, N, NTOTAL,
ALPHA=.05, POW=.8 ){

cat( "INPUT: FA=", FSTATA, "FB=", FSTATB, "FAB=", FSTATAB, "M=", M,
"N=", N, "ALPHA=", ALPHA, "POWER=", POW, "\n" )
cat( "\n" )

NCELLS <- M*N; phiE <- NTOTAL - NCELLS

phiA <- M-1; eta2A <- phiA*FSTATA/(phiA*FSTATA + phiE)
FHATA <- sqrt(eta2A/(1 - eta2A)); numA <- phiE/(phiA + 1) + 1
cat( "FhatA=", FHATA, "n_correctedA=", numA, "\n" )
ACPOWER <- pwr.anova.test( k=M, n=numA, f=FHATA, sig.level=ALPHA )$power
cat( "Achieved_power=", ACPOWER, "\n" )
cat( "\n" )

phiB <- N-1; eta2B <- phiB*FSTATB/(phiB*FSTATB + phiE)
FHATB <- sqrt(eta2B/(1 - eta2B)); numB <- phiE/(phiB + 1) + 1
cat( "FhatB=", FHATB, "n_correctedB=", numB, "\n" )
ACPOWER <- pwr.anova.test( k=N, n=numB, f=FHATB, sig.level=ALPHA )$power
cat( "Achieved_power=", ACPOWER, "\n" )
cat( "\n" )

if( is.na( FSTATAB ) == FALSE ){
phiAB <- (M-1)*(N-1);
eta2AB <- phiAB*FSTATAB/(phiAB*FSTATAB + phiE)
FHATAB <- sqrt(eta2AB/(1-eta2AB)); numAB <- phiE/(phiAB + 1) + 1
cat( "FhatAB=", FHATAB, "n_correctedAB=", numAB, "\n" )
ACPOWER <- pwr.anova.test( k=phiAB+1, n=numAB, f=FHATAB,
sig.level=ALPHA )$power
cat( "Achieved_power=", ACPOWER, "\n" )
cat( "\n" )
}
```

Fig. 7.9 Source code of `future.sample.2wayanova2` (top half)

N	The sample size n per group;
NTOTAL	The total number of observations, which, with r measurements per cell as shown in Table 3.4 (Chap. 3, Sect. 3.3), amounts to mnr;
ALPHA	α, the desired Type I error probability (default: 0.05);
POWER	$1 - \beta$, the desired statistical power (default: 0.80).

The code in Fig. 7.9 shows that `future.sample.2wayanova2` computes the achieved power using `pwr.anova.test` (from the `pwr` library) for Factor A (i.e. system effect), Factor B (i.e. topic effect), and the interaction $A \times B$. Let us take the system effect as an example. First, it can be observed that the partial effect size is computed as [4]:

$$\hat{\eta}_p^2 = \frac{\phi_A F_0}{\phi_A F_0 + \phi_{E3}} , \qquad (7.9)$$

where $\phi_A = m - 1$, $\phi_{E3} = mnr - mn = mn(r - 1)$ when there are r measurements per cell, mn is the total number of cells, and F_0 is the test statistic for two-way ANOVA with replication as given by Chap. 3, Sect. 3.3 Eqs. 3.42, 3.43, and 3.44, etc. Then, the partial effect size index is computed as:

```
pwr.anova.pre <- function( f, df, cell, power, sig.level ){
n <- pwr.anova.test( power=power, k=df+1, f=f,
sig.level=sig.level )$n
result <- (n-1)*(df+1)/cell + 1
cellsize<-t(ceiling(result))
return( cellsize )
}

PERCELLSIZ <- pwr.anova.pre( f=FHATA, df=phiA, cell=NCELLS, power=POW,
sig.level=ALPHA )
TOTALSIZ <- NCELLS*PERCELLSIZ
cat( "Sample size per cell (A) =", PERCELLSIZ, "\n" )
cat( "Sample size total (A) =", TOTALSIZ, "\n" )
cat( "\n" )

PERCELLSIZ <- pwr.anova.pre( f=FHATB, df=phiB, cell=NCELLS, power=POW,
sig.level=ALPHA )
TOTALSIZ <- NCELLS*PERCELLSIZ
cat( "Sample size per cell (B) =", PERCELLSIZ, "\n" )
cat( "Sample size total (B) =", TOTALSIZ, "\n" )
cat( "\n" )

if( is.na( FSTATAB ) == FALSE ){
PERCELLSIZ <- pwr.anova.pre( f=FHATAB, df=phiAB, cell=NCELLS,
power=POW, sig.level=ALPHA )
TOTALSIZ <- NCELLS*PERCELLSIZ
cat( "Sample size per cell (AB) =", PERCELLSIZ, "\n" )
cat( "Sample size total (AB) =", TOTALSIZ, "\n" )
cat( "\n" )
}

}
```

Fig. 7.10 Source code of `future.sample.2wayanova2` (bottom half)

$$\hat{f} = \sqrt{\frac{\hat{\eta}_p^2}{1 - \hat{\eta}_p^2}} \ . \tag{7.10}$$

It can also be observed from Fig. 7.9 that while `future.sample.2wayanova2` uses `pwr.anova.test` (from the `pwr` library) to compute the achieved power just like `future.sample.1wayanova` does, it feeds a *corrected* sample size per cell to `pwr.anova.test`:

$$\frac{\phi_{E3}}{\phi_A + 1} + 1 \ . \tag{7.11}$$

See Cohen [1] (p.365) or Toyoda [4] (p.196) for details.

The code in Fig. 7.10 first defines a function called `pwr.anova.pre` [4], in which `pwr.anova.test` is called first to obtain a sample size n for a factor (e.g. A) while ignoring the other factors (e.g. B and $A \times B$), and then the final estimate of the sample size per cell is computed as:

$$\frac{(n-1)(\phi+1)}{mn} + 1 \ , \tag{7.12}$$

```
> future.sample.2wayanova2( 24.00, 24.89, 10.03, 2, 2, 964 )
INPUT: FA= 24 FB= 24.89 FAB= 10.03 M= 2 N= 2 ALPHA= 0.05 POWER= 0.8

FhatA= 0.1581139 n_correctedA= 481
Achieved_power= 0.9983547

FhatB= 0.1610189 n_correctedB= 481
Achieved_power= 0.9987741

FhatAB= 0.1022151 n_correctedAB= 481
Achieved_power= 0.8863183

Sample size per cell (A) = 80
Sample size total (A) = 320

Sample size per cell (B) = 77
Sample size total (B) = 308

Sample size per cell (AB) = 189
Sample size total (AB) = 756
```

Fig. 7.11 Using `future.sample.2wayanova2`

where ϕ is either ϕ_A, ϕ_B or $\phi_{A \times B}$. See Cohen [1] (p.366) or Toyoda [4] (p.192) for details. Finally, it can be observed that `future.sample.2wayanova2` executes `pwr.anova.pre` for A (system effect), B (topic effect), and optionally for $A \times B$ (system-topic interaction) as well.

Figure 7.11 shows an example where $F_0 = 24$ for Factor A, $F_0 = 24.89$ for Factor B, and $F_0 = 10.03$ for $A \times B$, $m = 2$, $n = 2$, with the defaults used for the other arguments. The achieved powers are 99.8% for A, 99.9% for B, and 88.6% for $A \times B$, which are very high. Moreover, the required sample size per cell for achieving 80% power is 80 for A, 77 for B, and 189 for $A \times B$. Hence, if the researcher is interested in all three factors, the sample size per cell that should be used for achieving 80% power would be 189. Since the number of cells is $2 * 2 = 4$, the total number of measurements in the table would be $4 \times 189 = 756$.

7.8 Summary

Using the scripts discussed in this chapter, I demonstrated at SIGIR 2016 that some IR experiments are highly overpowered, while others are seriously underpowered [3]. Figure 7.12 shows two graphs copied from that work, where each dot represents the *sample size ratio* of a past SIGIR paper, computed as the ratio of the actual sample size used in the experiment to the desired sample size for achieving 80% statistical power. The R tools discussed in this chapter were used to plot the graph. For example, in the top graph (which concerns paired and two-sample t-tests), Paper #3 used a sample of size $n = 192$, when the sample size required for achieving 80% power was only $n' = 5$: a highly overpowered experiment. In contrast, in the same

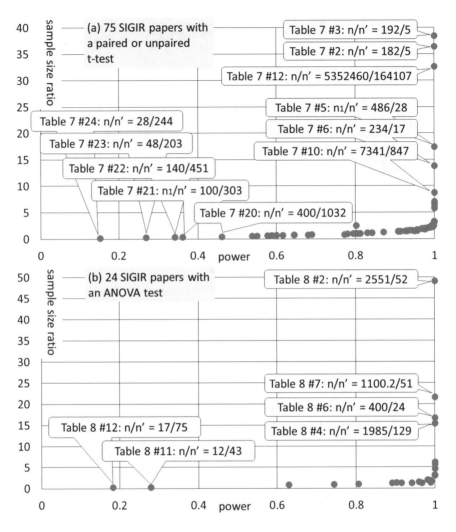

Fig. 7.12 Sample size ratio graphs from Sakai [3]

graph, Paper #24 used a sample of size $n = 28$, when the sample size required for achieving 80% power was $n' = 244$: a highly underpowered experiment.[3] The achieved power for Paper #3 was 100%, while that for Paper #24 was as low as 15.2%. See Sakai [3] for more details.

Highly overpowered experiments are often reported in papers from industry, where they deal with an abundance of data. Highly overpowered means that they are using far larger samples than necessary, which is not necessarily bad if they

---

[3]This is in fact the example we discussed in Sect. 7.3.

already have that much data and they are not wasting any resources for building such data from scratch for the experiment. However, the prerequisite is that they report the effect sizes, not just the p-values; as I have stressed in this book, large samples can always give you small p-values, even if the actual effect is small.

Highly *underpowered* experiments are clearly a problem. In IR, this often happens in a user study,[4] since it is very expensive to hire user study participants and this limits the sample sizes. A highly underpowered experiment means that the researcher is highly likely to miss true differences and therefore that she may be wasting resources for conducting the experiment. I recommend researchers to conduct a pilot study to first estimate the effect size for the factor of interest and then design a full experiment based on power considerations as we have discussed in this chapter.

References

1. J. Cohen, *Statistical Power Analysis for the Behavioral Sciences*, 2nd edn. (Psychology Press, New York, 1988)
2. D. Kelly, Methods for evaluating interactive information retrieval systems with users. Found. Trends Inf. Retr. **3**(1–2), 1–224 (2009)
3. T. Sakai, Statistical significance, power, and sample sizes: a systematic review of SIGIR and TOIS, in *Proceedings of ACM SIGIR*, Pisa, 2016, pp. 5–14
4. H. Toyoda, *Introduction to Statistical Power Analysis: A Tutorial with R (in Japanese)* (Tokyo Tosyo, Chiyoda, 2009)

---

[4]For general considerations required for designing user studies, see Kelly [2].

Chapter 8
Conclusions

Abstract This chapter first provides a quick summary of the topics covered in this book (Sect. 8.1). It then very briefly touches upon Bayesian approaches to hypothesis testing, not covered in the previous chapters, and concludes the book by proposing a statistical reform in IR (Sect. 8.2).

Keywords Bayesian statistics · Confidence intervals · Effect sizes · p-values · Statistical reform

8.1 A Quick Summary of the Book

In this book, I mainly focussed on classical parametric significance tests for comparing means, namely, t-tests, ANOVAs, and (randomised) Tukey HSD tests. I also discussed the criticisms on, and limitations of, classical statistical significance testing in general. The reader should be aware by now that there are four main factors in significance testing: the Type I error probability α, the Type II error probability β (where $(1 - \beta)$ represents the statistical power), the sample size (e.g. number of topics), and the effect size (e.g. magnitude of the difference between two IR systems). If you specify three of them, the fourth factor is determined automatically. I discussed topic set size design for test collection builders and power analysis for researchers who are planning to conduct experiments similar to those reported in the literature. Both of these methods utilise the relationship across the four factors. Test collections and experiments should not be designed blindly: we should start by estimating the variances and/or effect sizes from some pilot data or experiments.

While classical significance tests may harm research if used inappropriately, I believe that they can be useful tools if they are used appropriately with the right sample sizes and statistical power (as discussed in Chaps. 6 and 7) and if the test results are reported properly as I have exemplified in Chap. 5.

© Springer Nature Singapore Pte Ltd. 2018

T. Sakai, *Laboratory Experiments in Information Retrieval*,

The Information Retrieval Series 40, https://doi.org/10.1007/978-981-13-1199-4_8

8.2   Statistical Reform in IR?

It is not clear to me for how long the IR community will continue to embrace classical significance testing; IR researchers should be aware that Bayesian approaches to hypothesis testing are quickly gaining popularity among statisticians and among research disciplines that have heavily used statistics [4, 7, 8]; we are beginning to see Bayesian approaches in the field of IR evaluation as well [1, 6, 9]. While the p-value is the probability of observing the observed data or something more extreme *under the null hypothesis*, Bayesian statistics can give us the probability that your research hypothesis is correct *given the data*. However, as long as IR researchers are content with the classical t-tests, ANOVA, (randomised) Tukey HSD, and the like, I believe that the recommendations discussed in this book will remain valid.

How can the entire IR community change for the better based on the recommendations provided in this book? I propose that IR journal editors and conference programme chairs incorporate these viewpoints (i.e. sample sizes and power considerations and reporting results informatively) in their review criteria and announce this change to the community. In particular, it is probably worth repeating that p-values are more informative than the dichotomous discussion of significant vs. not significant; effect sizes and confidence intervals are more informative than p-values. IR research deserves a *statistical reform* [2, 3, 5].

References

1. B. Carterette, Bayesian inference for information retrieval evaluation, in *Proceedings of ACM ICTIR*, Northampton, 2015, pp. 31–40
2. F. Fidler, G. Cumming, Lessons learned from statistical reform efforts in other disciplines. Psychol. Sch. **44**(5), 441–449 (2007)
3. F. Fidler, C. Geoff, B. Mark, T. Neil, Statistical reform in medicine, psychology and ecology. J. Socio-Econ. **33**, 615–630 (2004)
4. J.K. Kruschke, *Doing Bayesian Data Analysis*, 2nd edn. (Elsevier, Amsterdam, 2015)
5. T. Sakai, Statistical reform in information retrieval? SIGIR Forum **48**(1), 3–12 (2014)
6. T. Sakai, The probability that your hypothesis is correct, credible intervals, and effect sizes for IR evaluation, in *Proceedings of ACM SIGIR*, Shinjuku, 2017, pp. 25–34
7. H. Toyoda (ed.), *Fundamentals of Bayesian statistics: Practical Getting Started by Hamiltonian Monte Carlo Method (in Japanese)* (Asakura Shoten, Shinjuku, 2015)
8. H. Toyoda, *An Introduction to Statistical Data Analysis: Bayesian Statistics for 'post p-value era'* (Asakuha Shoten, Shinjuku, 2016)
9. D. Zhang, J. Wang, E. Yilmaz, X. Wang, Y. Zhou, Bayesian performance comparison of text classifiers, in *Proceedings of ACM SIGIR*, Pisa, 2016, pp. 15–24

Index

A
Alternative hypothesis, 4, 29, 45, 68, 104
Analysis of variance, 43, 89
Average precision, 2, 121

B
Bayesian hypothesis testing, 148
Beta function, 13, 16, 23
Bonferroni correction, 61
Bootstrap test, 28, 74

C
Central limit theorem, 9, 27
Chi-square distribution, 11
Cohen's d, 86
Cohen's effect size index, 138
Cohen's five-eighty convention, 6, 112
Comparisonwise error rate, 61
Confidence interval, 39, 56, 83, 93, 95, 115, 118
Contrast, 60
Critical value, 4

D
Degrees of freedom, 11, 13, 15, 16, 32
Dichotomous thinking, 84

E
Effect size, 85, 93, 99, 134
Expectation, 6
Expected reciprocal rank, 121

F
Familywise error rate, 60
F distribution, 16, 47

G
Gamma function, 12, 18, 22, 87, 117
Glass' delta, 87

H
Hedge's g, 86, 137
Homoscedasticity, 30, 45, 73, 86, 88, 94

I
Independence, 7
Interaction, 51, 91, 140

L
Law of large numbers, 7

M
Margin of error, 39, 56, 71, 116
Minimum detectable difference, 103, 108
Minimum detectable range, 111

N
Noncentral chi-square distribution, 22
Noncentral F distribution, 23, 113
Noncentrality parameter, 17, 22, 23, 105, 113
Noncentral t distribution, 17, 105, 109
Normal distribution, 8, 27
Normalised discounted cumulative gain, 2, 27, 121

© Springer Nature Singapore Pte Ltd. 2018
T. Sakai, *Laboratory Experiments in Information Retrieval*,
The Information Retrieval Series 40, https://doi.org/10.1007/978-981-13-1199-4

Null distribution, 4
Null hypothesis, 4, 29, 45, 68

O
One-sided test, 4
One-way analysis of variance, 44, 89, 111, 137
One-way analysis of variance with equal group
    sizes, 44
One-way analysis of variance with unequal
    group sizes, 47
Overpowered experiment, 133, 143

P
Paired t-test, 29, 62, 93, 102, 134
Parametric test, 1, 28
Partial effect size, 90, 140
Pooled variance, 15, 31, 87
Population correlation coefficient, 88
Population mean, 2, 6, 45, 56
Population standard deviation, 7, 86
Population variance, 7, 86, 89, 120
Probability density function, 3, 8, 12, 13, 16,
    18, 22, 23
p-value, 4, 75, 78, 83, 93

Q
Q-measure, 121

R
Randomisation test, 28, 74
Randomised Tukey's honestly significant
    difference test, 77, 88, 95
Random sample, 2, 73
Random variable, 2
Reproductive property, 9, 12

S
Sample mean, 2, 7, 9
Sample size, 6, 7, 32, 84, 99, 134

Sample size design, 99
Sample variance, 10, 17, 32
Significance criterion, 4, 84
Simultaneous confidence interval, 71
Single-step method, 59
Standardisation, 8
Standardised mean difference, 84, 85, 135
Standard normal distribution, 8
Statistical power, 6, 99, 133
Statistical reform, 148
Statistical significance, 5, 81
Studentised range distribution, 68
Student's two-sample t-test, 30, 32, 93, 108,
    136
Sum of squares, 10, 12, 45, 50, 52

T
t distribution, 13, 29
Test statistic, 4, 84
Text Retrieval Conference, 100, 121
Topic set size design, 99
Tukey's honestly significant difference test, 67,
    88
Two-sample t-test, 30, 32, 93, 108, 136
Two-sided test, 4
Two-way analysis of variance without
    replication, 49, 139
Two-way analysis of variance with replication,
    51, 90, 91, 140
Type I error, 5, 61, 99
Type II error, 5, 99

U
Unbiased estimator, 10, 29, 86–88
Underpowered experiment, 133, 143

W
Welch's two-sample t-test, 31

Printed in the United States
By Bookmasters